T0257543

Industrial Aspects of Finite Element Analysis

Industrial Aspects of Finite Element Analysis

Edited by **Connie Mcguire**

New York

Published by NY Research Press,
23 West, 55th Street, Suite 816,
New York, NY 10019, USA
www.nyresearchpress.com

Industrial Aspects of Finite Element Analysis
Edited by Connie Mcguire

© 2015 NY Research Press

International Standard Book Number: 978-1-63238-292-4 (Hardback)

Printed in the United States of America.

Contents

Preface

Finding approximate solutions to partial differential equations, and integral equations, allowing numerical assessment of complicated structures based on their material properties is best represented by the mathematical method of Finite Element Analysis. This book presents varied topics on the utilization of Finite Elements, varying from manufacturing technology to industrial developments. The structure and language of the book has been so written that it is useful for graduate students learning 'applications of finite element modeling and analysis' and also encompasses topics and reference material useful for research and professionals who want to gain a deeper knowledge of finite element analysis.

All of the data presented henceforth, was collaborated in the wake of recent advancements in the field. The aim of this book is to present the diversified developments from across the globe in a comprehensible manner. The opinions expressed in each chapter belong solely to the contributing authors. Their interpretations of the topics are the integral part of this book, which I have carefully compiled for a better understanding of the readers.

At the end, I would like to thank all those who dedicated their time and efforts for the successful completion of this book. I also wish to convey my gratitude towards my friends and family who supported me at every step.

Editor

Materials, Structures, Manufacturing Industry and Industrial Developments

Contact Stiffness Study: Modelling and Identification

Hui Wang[1], Yi Zheng[2] and Yiming (Kevin) Rong[1,2]
[1]Tsinghua University,
[2]Worcester Polytechnic Institute,
[1]China
[2]USA

1. Introduction

In machining processes, fixtures are used to accurately position and constrain workpiece relative to the cutting tool. As an important aspect of tooling, fixturing significantly contributes to the quality, cost, and the cycle time of production. Fixturing accuracy and reliability is crucial to the success of machining operations.

Computerized fixture design &analysis has become means of providing solutions in production operation improvement. Although fixtures can be designed by using CAD functions, a lack of scientific tool and systematic approach for evaluating the design performance makes them rely on trial-and-errors, which leads to several problems, for instance, over design in functions, which is very common and sometimes depredates the performance (e.g., unnecessary heavy design); the quality of design that cannot be ensured before testing; the long cycle time of fixture design, fabrication, and testing, which may take weeks if not months; a lack of technical evaluation of fixture design in the production planning stage.

Over past two decades, Computerized Aided Fixture Design (CAFD) has been recognized as an important area and studied from fixture planning, fixture design to fixturing analysis/verification. The fixture planning is to determine the locating datum surfaces and locating/clamping positions on the workpiece surfaces for a totally constrained locating and reliable clamping. The fixture design is to generate a design of fixture structure as an assembly, according to different production requirements such as production volume and machining conditions. And the design verification is to evaluate fixture design performances for satisfying the production requirements, such as completeness of locating, tolerance stack-up, accessibility, fixturing stability, and the easiness of operation.

For many years, fixture planning has been the focus of fixture related academic research with significant progress in both theoretical and practical studies. Most analyses are based on strong assumptions, e.g., frictionless smooth surfaces in contact, rigid fixture body, and single objective function for optimization. Fixture design is a complex problem with considerations of many operational requirements. Four generations of CAFD techniques and systems have been developed: group technology (GT)-based part classification for fixture design and on-screen editing, automated modular fixture design, permanent fixture design with predefined fixture components types, and variation fixture design for part families. The study on a new generation of CAFD just started to consider operational requirements. Both geometric reasoning, knowledge-based as well as case-based reasoning (CBR) techniques have been intensively studied for CAFD. How to make use of the best practice knowledge in fixture design and verify the fixture design quality under different conditions has become a challenge in the fixture design &analysis study.

In fixture design verification, it was proved that when the fixture stiffness and machining force are known as input information, the fixturing stability problem could be completely solved. However most of the studies were focused on the fixtured workpiece model, i.e., how to configure positions of locators and clamps for an accurate and secured fixturing. FEA method has been extensively used to develop fixtured workpiece model (e.g., Fang, 2002; Lee, 1987; Trappey, 1995) with an assumption of rigid or linear elastic fixture stiffness. The models and computational results cannot represent the nonlinear deformation in fixture connections identified in previous experiments. As Beards (1983) pointed out, up to 60% of the deformation and 90% of the damping in a fabricated structure can arise from various connections. The determination of fixture contact stiffness is the key barrier in the analysis of fixture stiffness. The existing work is very preliminary, by either simply applying the Hertzian contact model or considering the effective contact area.

The development of fixture design &analysis tools would enhance both the flexibility and the performance of the workholding systems by providing a more systematic and analytic approach to fixture design. Fixture stationary elements, such as locating pads, buttons, and pins, immediately contact with the workpiece when loading the workpiece. Subsequent clamping (by moveable elements) creates pre-loaded joints between the workpiece and each fixture component. Besides, there may be supporting components and a fixture base in a fixture. In fixture design, a thoughtful, economic fixture-workpiece system maintains uniform maximum joint stiffness throughout machining while also providing the fewest fixture components, open workpiece cutting access, and shortest setup and unloading cycles. Both static and dynamic stiffness in this fixture-workpiece system rely upon the component number, layout and static stiffness of the fixture structure. These affect fixture performance and must be addressed through appropriate design solutions integrating the fixture with other process elements to produce a highly rigid system. This requires a fundamental understanding of the fixture stiffness in order to develop an accurate model of the fixture - workpiece system.

2. Computer-aided fixture design with predictable fixture stiffness

The research on fixture-workpeice stiffness is a crucial topic in fixture design field. Currently, based on the elastic body assumption, using FEA method to predict the fixture stiffness has been widely accepted. With the consideration on the contact and friction conditions, the validity and accuracy of the methodology was been illustrated by two cases simulation and experimental comparison (Zhu, 1993).

The following is an introduction on the general methodology.

First the stiffness of typical fixture units is studied with considerations of contact friction conditions. The results of the fixture unit stiffness analysis are integrated in fixture design as a database with variation capability driven by parametric representations of fixture units. When a fixture is designed using fixture design &analysis tool, the fixture stiffness at the contact locations (locating and clamping positions) to the workpiece can be estimated and/or designed based on the machining operation constraints (e.g., fixture deformation and dynamic constraints). Fig. 1 shows a diagram of the integrated fixture design system.

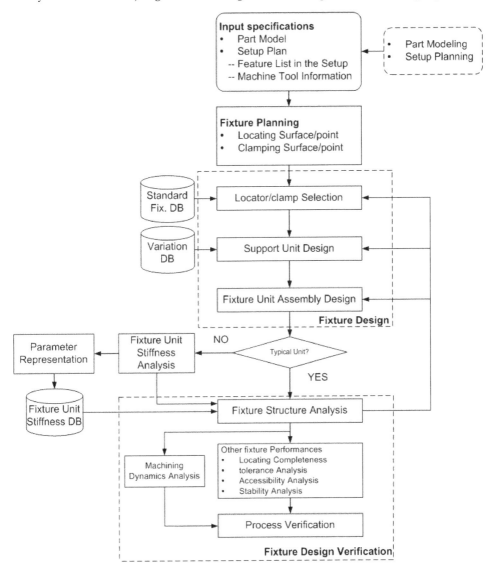

Fig. 1. Integrated Fixture Design System

In order to study the fixture stiffness in a general manner, fixture structure is decomposed into functional units with fixture components and functional surfaces (Rong, 1999). In a fixture unit, all components are connected one to another where only one is in contact directly with fixture base and one or more in contact with the workpiece serving as the locator, clamp, or support. Fig. 2 shows a sketch of the fixture units in a fixture design. When a workpiece was located and clamped in the fixture, the fixture units are subjected to the external loads that pass through the workpiece. If the external load is known and acting on a fixture unit, and the displacement of the fixture unit at the contact position is measured or calculated based on a finite element (FE) model, the fixture unit stiffness can be determined.

The fixture unit stiffness is defined as the force required for a unit deformation of the fixture unit in normal and tangential directions at the contact position with workpiece. The stiffness can be static if the external load is static (such as clamping force), and dynamic if the external load is dynamic (such as machining force). It is the key parameter to analyze the relative performance of different fixture designs and optimize the fixture configuration.

Analysis of fixture unit stiffness may be divided into three categories: analytical, experimental and finite element analysis (FEA). Conventional structural analysis methods may not work well in estimating the fixture unit stiffness. Preliminary experimental study has shown the nature of fixture deformation in T-slot based modular fixtures (Zhu, 1993). An integrated model of a fixture-workpiece system was established for surface quality prediction (Liao, 2001) based on the experiment results in (Zhu, 1993), but combining zhu's experimental work and finite element analysis (FEA). Hurtado used one torsional spring, and two linear springs, one in the normal direction and the other in the tangential direction, to model the stiffness of the workpiece, contact and fixture element. (Hurtado, 2002) FEA method has not been studied for fixture unit stiffness due to the complexity of the contact conditions and the large computation effort for many fixture components involved.

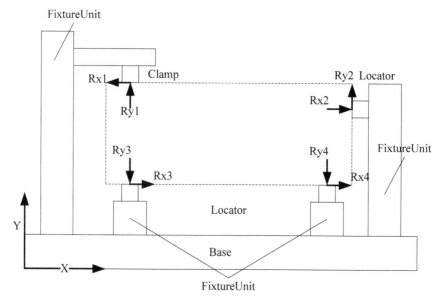

Fig. 2. Sketch of Fixture Units

3. Finite element model with frictional contact conditions

3.1 Finite element formulation

Consider a general fixture unit with two components I and J, as shown in Fig.3 (Zheng, 2005, 2008). For multi-component fixture units, the model can be expanded. The fixture unit is discretized into finite element models using a standard procedure, except for the contact surfaces, where each nodes on the finite element mesh for the contact surface is modelled by a pair of nodes at the same location belonging to components I and J, respectively, which are connected by a set of contact elements. The basic assumptions include that material is homogenous and linearly elastic, displacements and strains are small in both components I and J, and the frictional force acting on the contact surface follows the Coulomb law of friction.

The total potential energy p of a structural element is expressed as the sum of the internal strain energy U and the potential energy Ω of the nodal force; that is,

$$\prod p = U + \Omega \tag{1}$$

It is well known that the element strain energy can be expressed as,

$$U = \frac{1}{2}\{q\}^{T}[K]\{q\} \tag{2}$$

where $[K]$ is the element stiffness matrix; and $\{q\}$ is the element nodal displacement vector.

The potential energy of the nodal force is,

$$\Omega = -\{q\}^{T}\{R\} \tag{3}$$

where $\{R\}$ is the vector of the nodal force. It includes internal force and external force.

When the two components I and J are in contact, a number of three-dimensional contact elements are in effect on the contact surfaces. It should note that the problem is strongly nonlinear, partially due to the fact that the number of contact elements may vary with the change of contact condition. The original contacting nodes might separate or recontact after separation, based on the deformation condition on the contact surface; also contact stiffness may not constant either. The contact elements are capable of supporting a compressive load in the normal direction and tangential forces in the tangential directions. When the two components are in contact, and the displacements in the tangential directions and normal direction are assumed as independent, the element itself can be treated as three independent contact springs: two having stiffness k_t and k in the tangential directions of the contact surface at the contact point and one having stiffness k_n in the normal direction.

Usually, there are two methods used to include the contact condition in the energy equation: the Lagrange multiplier and the penalty function methods. In order to understand these methods, a physical model of the contact conditions is presented, shown in Fig. 4. When two contact surfaces of fixture components, i.e., body J and I, are loaded together, they will contact at a few asperities.

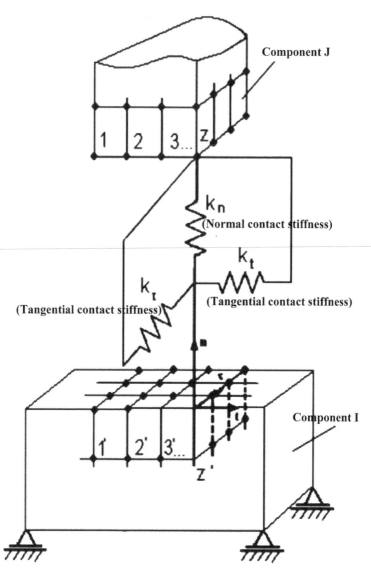

Fig. 3. Contact Model of Two Fixture Components

The contact criteria can be written as:

$$\eta \geq 0; \quad f_{ni} \leq 0; \quad \eta f_{ni} = 0$$

Where,
η is distance from a contact point i in body I to a contact point j on the body J in the normal direction of contact; f_{ni} is the contact force acting on point i of body I in the normal direction.

It shows the kinematic condition of no penetration and the static condition of compressive normal force. To prevent interpenetration, the separation distance η for each contact pair must be greater or equal to zero. If $\eta>0$, the contact force $f_{ni}=0$. When $\eta=0$, the points are in contact and $f_{ni}<0$. If $\eta<0$, penetration occurs. In real physics, the actual contact area increases, and contact stiffness is enhanced when the load increases. Therefore, the contact deformation is nonlinear as a function of the preload as shown Figure 4(e). In the Lagrange multiplier method, the function $w(\eta, f_{ni})$ represents the constraint, which prevents the penetration between contact pairs. In the penalty function method, an artificial penalty parameter is used to prevent the penetration between contact pairs.

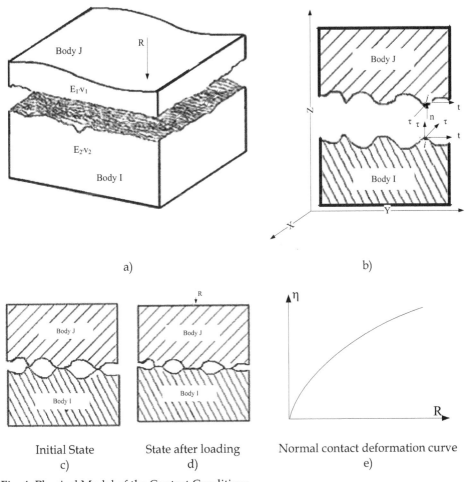

a)   b)

Initial State   State after loading   Normal contact deformation curve
c)   d)   e)

Fig. 4. Physical Model of the Contact Conditions

In the penalty function method, the contact condition is represented by the constraint equation,

$$\{t\} = [K_C]\{q\} - \{Q\} \tag{4}$$

Where $\{t\}$ is the constraint equation, $[K_C]$ is the contact element stiffness matrix, $\{Q\}$ is the contact force vector of the active contact node pairs. When $\{t\} = \{0\}$, it means that the constraints are satisfied. So the constraint equation Eq. 4 becomes

$$[K_C]\{q\} = \{Q\} \tag{5}$$

The total potential energy Π_p in Eq. 1 can be augmented by a penalty function $\frac{1}{2}\{t\}^T[\alpha]\{t\}$ where $[\alpha]$ is a diagonal matrix of penalty value α_i. The total potential energy in the penalty function method becomes

$$\Pi_{pP} = \frac{1}{2}\{q\}^T[K]\{q\} - \{q\}^T\{R\} + \frac{1}{2}\{t\}^T[\alpha]\{t\} \tag{6}$$

The minimization of Π_{pP} with respect to $\{q\}$ requires that $\left\{\frac{\partial \Pi_{pP}}{\partial q}\right\} = \{0\}$, which leads to

$$\left([K] + [K_C]^T[\alpha][K_C]\right)\{q\} = \{R\} + [K_C]^T[\alpha]\{Q\} \tag{7}$$

where $[K_C]^T[\alpha][K_C]$ is the penalty matrix.

On the other hand, in the Lagrange multiplier method, the contact constraint equation can be written as:

$$w = \{\eta\}^T\left([K_C]\{q\} - \{Q\}\right) \tag{8}$$

where the components of the row vector η_i (i=1, 2, ..., N), are often defined as Lagrange multipliers η_i.

Adding Eq. 8 to the potential energy in Eq. 1, we have the total energy in the Lagrange multiplier method,

$$\Pi_{pL} = \frac{1}{2}\{q\}^T[K]\{q\} - \{q\}^T\{R\} + \{\eta\}^T\left([K_C]\{q\} - \{Q\}\right) \tag{9}$$

The minimization of Π_{pL} with respect to $\{q\}$ and $\{\eta\}$ requires that $\left\{\frac{\partial \Pi_{pL}}{\partial q}\right\} = \{0\}$ and $\left\{\frac{\partial \Pi_{pL}}{\partial \eta}\right\} = \{0\}$, which leads to,

$$\left\{\frac{\partial \Pi_{pL}}{\partial q}\right\} = [K]\{q\} + [K_C]^T\{\eta\} - \{R\} = \{0\} \tag{10}$$

$$\left\{ \frac{\partial \Pi_{pL}}{\partial \eta} \right\} = [K_C]\{q\} - \{Q\} = \{0\} \tag{11}$$

In a matrix form, Eqs. 10 and 11 can be expressed as,

$$\begin{bmatrix} [K] & [K_C]^T \\ [K_C] & [0] \end{bmatrix} \begin{Bmatrix} \{q\} \\ \{\eta\} \end{Bmatrix} = \begin{Bmatrix} \{R\} \\ \{Q\} \end{Bmatrix} \tag{12}$$

While the constraints in Eq. 8 can be satisfied, the Lagrange multiplier method has disadvantages. Because the stiffness matrix in Eq. 12 may contain a zero component in its diagonal, there is no guarantee of the absence of the saddle point. In this situation, the computational stability problem may occur. In order to overcome that difficulty, a perturbed Lagrange multiplier method was introduced (Aliabadi, 1993).

$$\begin{aligned} \Pi^p_{pL} &= \Pi_{pL} - \frac{1}{2\alpha'}\{\eta\}^T\{\eta\} \\ &= \frac{1}{2}\{q\}^T[K]\{q\} - \{q\}^T\{R\} + \{\eta\}^T\left([K_C]\{q\} - \{Q\}\right) - \frac{1}{2\alpha'}\{\eta\}^T\{\eta\} \end{aligned} \tag{13}$$

where α' is an arbitrary positive number. At the limit α' goes to ∞, the perturbed solutions converge to the original solutions. The introduction of α' will maintain a small force across and along the interface. This will not only maintain stability but also avoid the stiffness matrix being singular, due to rigid body motion. Similarly, the minimization of Π^p_{pL} with respect to $\{q\}$ and $\{\eta\}$ results in the following matrix,

$$\begin{bmatrix} [K] & [K_C]^T \\ [K_C] & -\frac{1}{\alpha'}[I] \end{bmatrix} \begin{Bmatrix} \{q\} \\ \{\eta\} \end{Bmatrix} = \begin{Bmatrix} \{R\} \\ \{Q\} \end{Bmatrix} \tag{14}$$

Eq. 14 can be expressed as:

$$[K]\{q\} = \{R\} - [K_C]^T\{\eta\} \tag{15}$$

$$\{\eta\} = \alpha'\left([K_C]\{q\} - \{Q\}\right) \tag{16}$$

Substitute Eq. 16 into Eq. 15,

$$\left([K] + [K_C]^T \alpha'[K_C]\right)\{q\} = \{R\} + [K_C]^T \alpha'\{Q\}$$

For simplicity, let all αi in $[\alpha]$ of penalty function equal to α', i.e. $\alpha_i = \alpha'$. Thus, the perturbed Lagrange multiplier is equivalent to the penalty function method.

In the Lagrange multiplier method, both displacement and contact force are regarded as independent variables; thus, the constraint (contact) conditions can be satisfied and the contact force can be calculated. It has disadvantages. The stiffness matrix contains zero components in its diagonal, and the Lagrange multiplier terms must be treated as additional variables. This leads to the construction of an augmented stiffness matrix, the order of which may significantly exceed the size of the original problem in the absence of constraint equations (Aliabadi, 1993). In comparison with the Lagrange multipliers method, the implementation of the penalty function method is relatively simple and does not require additional independent variables. It is often adopted in the practical analysis because of its simple implementation.

3.2 Contact conditions

Based on an iterative scheme (Mazurkiewicz, 1983), the contact conditions in FEA model are classified into the following three cases:

1. Open condition: gap remains open;
2. Stick condition: gap remains closed, and no sliding motion occurs in the tangential directions; and
3. Sliding condition: gap remains closed, and the sliding occurs in the tangential directions.

Let f_{ji} and u_{ji} be the contact nodal load vector and the nodal displacement, respectively, which are defined in the local coordinate system, where the subscript j indicates the component number ($j = I$ or J), and i indicates the coordinate ($i = n, t, \tau$), as shown in Fig. 5. By equilibrium of the contact element, $\vec{f}_{In} + \vec{f}_{It} + \vec{f}_{I\tau} + \vec{f}_{Jn} + \vec{f}_{Jt} + \vec{f}_{J\tau} = 0$. F_i ($i = n, t, \tau$) is the external nodal load in i direction $\{R\} = \sum_{x=1}^{n} \begin{pmatrix} F_n \\ F_t \\ F_\tau \end{pmatrix}_x$ where x is the node number of body I or J.

The displacement and force must satisfy the equilibrium equations in the three contact conditions (*note that {n, t, τ} is the local coordinate system*).

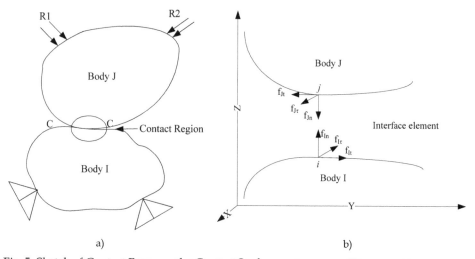

a)   b)

Fig. 5. Sketch of Contact Force on the Contact Surface

3.2.1 Open condition

When the normal nodal force F_n is positive (tension), the contact is broken, and no force is transmitted. The displacement change in the normal and tangential directions, denoted respectively by Δu_i $(i = n, t, \tau)$, then is

$$\Delta u_n = u_{Jn} - u_{In} + \delta_n, \quad f_{Ji} = f_{Ii} = 0 \quad (i = n, t, \tau) \tag{17}$$

where u_{Jn} and u_{In} are the current displacements of node J and node I in a normal direction, respectively. For each structural contact element, stiffness and forces are updated, based upon current displacement values, in order to predict new displacements and contact forces. δ_n is the gap between a pair of the potential contact points. In each increment of load, the gap status and the stiffness values are iteratively changed until convergence. As the load is increased, δ_n will change and hence should be adjusted as $\delta_n = \delta^0{}_n - \delta^T{}_n$, where $\delta^0{}_n$ is the initial gap before any deformation and $\delta^T{}_n$ is the gap change caused by the total combined normal movement at the pair of points.

3.2.2 Stick condition

The force in the tangential direction (F_S), which is the composition of the nodal force in t and τ directions $(F_t$ and $F_\tau)$, is defined only when $F_n < 0$ (compression). When the absolute value of F_S is less than $\mu | F_n |$, where μ is the Coulomb friction coefficient, there is no slide-motion in the interface, and the contact element responds like a spring. The stick condition exists if $\mu | F_n | > \left\| \left(u_{Jt} k_t + u_{J\tau} k_\tau \right) - \left(u_{It} k_t + u_{I\tau} k_\tau \right) \right\|$. That is,

$$f_{Ii} = -f_{Ji}, \quad u_{Jn} - u_{In} + \delta_n = 0, \quad u_{Ji} - u_{Ii} = 0, \quad (i = t\ \tau) \tag{18}$$

where k_t and k_τ are the tangential contact stiffness in t and τ directions, respectively. In the analysis of fixture unite stiffness, set $k_t = k_\tau$.

3.2.3 Sliding condition

Slide-motion will occur when the absolute value of F_S is more than $\mu | F_n |$. The slide-motion may occur in both the element t and τ directions. That is, if $\mu | F_n | < \left\| \left(u_{Jt} k_t + u_{J\tau} k_\tau \right) - \left(u_{It} k_t + u_{I\tau} k_\tau \right) \right\|$, then,

$$f_{It} = -f_{Jt} = \left(\pm \mu F_n \right)_t, \quad f_{I\tau} = -f_{J\tau} = \left(\pm \mu F_n \right)_\tau, \quad f_{In} = -f_{Jn}, \quad u_{In} - u_{Jn} + \delta_n = 0 \tag{19}$$

where $\left(\pm \mu F_n \right)_t$ and $\left(\pm \mu F_n \right)_\tau$ mean the maximum friction force in t and τ directions.

3.3 Solution procedure

The model presented in the previous section can be implemented to determine the fixture unit stiffness in clamping and machining. Because the model involves high nonlinearity, the

Newton-Raphson (*N-R*) approach is used to solve the problem. Considering the full Newton-Raphson iteration it is recognized that in general the major computational cost per iteration lies in the calculation and factorization of the stiffness matrix. Since these calculations can be quite expensive when large-order systems are considered, the modified Newton-Raphson algorithm is used in this research (Bathe, 1996).

Given the applied load R and the corresponding displacement u, the applied load is divided into a series of load increments. At each load step, the contact stiffness and contact conditions remain constant. And several iterations may be necessary to find a solution with acceptable accuracy. The modified Newton-Raphson method is used first to evaluate the initial out-of-balance load vector at the beginning of the iteration at each load step. The out-of-balance load vector is defined as the difference between the applied load vector R and the vector of restoring loads R_i^r. When the out-of-balance load is non-zero, the program performs a linear solution, using the initial out-of-balance loads, and then checks for convergence. If the convergence criteria are not satisfied, the out-of-balance load vector is reevaluated, the new contact conditions and the stiffness matrix are updated, and a new solution is obtained. This iterative procedure continues until the solution converges. The modified Newton-Raphson method and its flowchart are outlined by Fig.6.

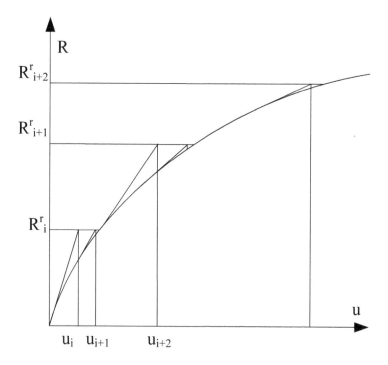

Fig. 6. (a) Modified Newton-Raphson Method

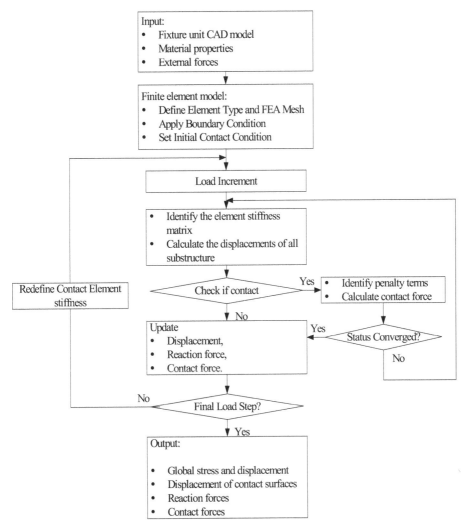

Fig. 6. (b) Flow Chart of the Analysis Procedure

4. Contact stiffness identification using a dynamic approach

First, the dynamic method is studied for use in the estimation of normal contact stiffness. The results of the dynamic methods are compared with the results based on the static test of normal contact stiffness; then the validated dynamic test method is used in estimation of tangential contact stiffness.

4.1 Theoretical formulation of 1-D normal contact stiffness

The idea behind the identification of normal contact stiffness is that the contact interface is modeled by a discrete linear spring. When the preload is changed, contact stiffness will change. When body *I* is in contact with the ground, the dynamic model of the entire structure can be shown as in Fig.7. According to this theoretical model the relationship between natural frequencies and normal contact stiffness can be established. When natural frequencies are obtained from impact test, along with a theoretical model, normal contact stiffness can be estimated.

In the one-dimensional model of body *I*, m is the mass of body *I*, k is the contact stiffness, *p* is the preload, *f(t)* is impulse excitation, *u(x,t)* is the longitudinal displacement of the bar at distance x from a fixed reference.

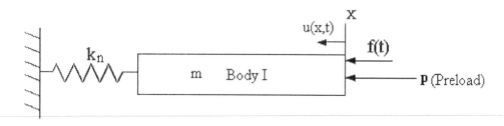

Fig. 7. One-Dimensional Model for Normal contact Stiffness

With use of a bar in Fig.7, the governing equation of the longitudinal vibration of the bar can be expressed as

$$\rho A \frac{\partial^2 u(x,t)}{\partial t^2} = \frac{\partial}{\partial x}\left(EA\frac{\partial u(x,t)}{\partial x}\right) \tag{20}$$

The boundary conditions of the bar are:

At *x=0*:
$$EA\frac{\partial u(0,t)}{\partial x} = 0 \tag{21}$$

and at *x=l*:
$$EA\frac{\partial u(l,t)}{\partial x} = -k_n u \tag{22}$$

Initially, the system starts from rest, from the static equilibrium position of the bar, such that the initial displacement condition is:

At *t=0*,
$$u(x,0) = 0 \tag{23}$$

The response of a system to an impulsive force can also be obtained by considering that the impulse produces an instantaneous change in the momentum of the system before any appreciable displacement occurs. The second initial condition is

$$\frac{\partial u(x,0)}{\partial t} = \frac{1}{m} \tag{24}$$

Assume
$$u(x,t) = X(x)q(t) \tag{25}$$

Substitute Eq. 25 into Eq. 20 to obtain

$$X(x)\frac{d^2q(t)}{dt^2} = c^2\frac{d^2X(x)}{dx^2}q(t)$$

$$\frac{1}{X(x)}\frac{d^2X(x)}{dx^2} = \frac{1}{c^2q(t)}\frac{d^2q(t)}{dt^2} = -\lambda^2$$

(26)

$-\lambda^2$ is called the separation constant and is designated to be negative (De Silva, 1999). Therefore, the mode shapes $X(x)$ satisfies

$$\frac{d^2X(x)}{dx^2} + \lambda^2X(x) = 0$$

(27)

whose general solution is $X(x) = C_1\sin\lambda x + C_2\cos\lambda x$ (28)

According to the general solution and the modal boundary conditions, one can get

$$\tan\lambda l = \frac{k_n}{EA\lambda}$$

(29)

Set the structure stiffness as $k^* = \dfrac{EA}{l}$ and the ratio of the stiffness as $\beta = \dfrac{k_n}{k^*}$. Since the structure stiffness k* is constant and known, the ratio of the stiffness β is proportional to the contact stiffness k_n. Therefore Eq. 11 can be expressed as

$$\tan\lambda l = \frac{\dfrac{k_n}{k^*}}{\lambda l} = \frac{\beta}{\lambda l}$$

(30)

This transcendental equation has an infinite number of solutions λ_i (i=1,2,...)that correspond to the modes of vibration. When β is changed, the solution of λi will change. When β is changed from 0.1 to 10, one can get the corresponding $\lambda_i l$ as shown in Fig.8. The natural frequencies can also be obtained using

$$\omega_i = \lambda_i c = \lambda_i\sqrt{\frac{E}{\rho}}$$

(31)

In an experimental study, the natural frequencies can be obtained by an impact test. λ_i can calculated from Eq.31 since the natural frequencies are related to the system characteristics. Then β can be determined from Eq. 30. Finally, the contact stiffness, k_n, can be estimated based on the definition of β. According to the assumption that contact stiffness is a function of the preload, the natural frequencies can be determined in experiments under different preloads. The change of contact stiffness can then be identified based on the change of the preloads, through measurement of the natural frequency variation. It should be noted that although any mode of the natural frequency can be used to estimate the contact stiffness, some modes might be more sensitive than others to the change of the preloads.

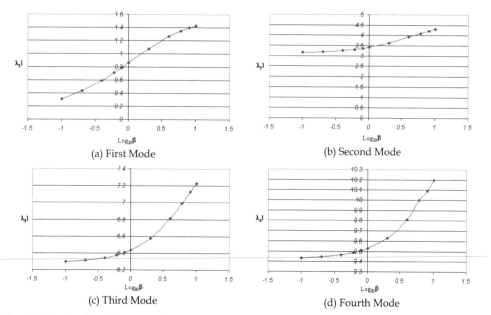

Fig. 8. Relationships between the Nondimensional Natural Frequencies and the Stiffness Ratio β in the First Four Modes

4.2 Experimental procedure and results

The experiments were conducted in order to verify the method of identifying contact stiffness in the normal direction (Zheng, 2005, 2008). The measurement instrumentation includes the proximity, the impact hammer with a load cell, power supply, and a Fast Fourier transformation (FFT) analyzer, as shown in Fig.9. The experimental procedure can be expressed as follows:

1. Frequency response function (FRF) of the bar is measured by using the hammer to excite the system. Thus, the natural frequencies of the bar can be obtained.

2. According to the natural frequency equation $\varpi_i = \lambda_i \sqrt{\dfrac{E}{\rho}}$, λ_i is calculated.

3. Based on the relationship between $\lambda_i l$ and β of the first three modes in Fig.8, the β can be inferred from the comparison of experimental results and theoretical results. Then the normal contact stiffness can be obtained from the equation $\beta = \dfrac{k_n}{k^*}$.

When the natural frequencies are obtained from the experiment, along with the curves of the relationships between $\lambda_i l$ and β, contact stiffness can be determined from each mode of vibration. However, when the preload changes, the natural frequencies may not necessarily change significantly with the change of normal load for certain modes. Contact stiffness should be identified from the mode most sensitive to changes of a preload. Fig.10 shows the FRF of the test system under different preload. Fig.11 shows the relationships between the natural frequencies and preload. The natural frequency of the third mode $f3$ is the most sensitive to changes in a preload.

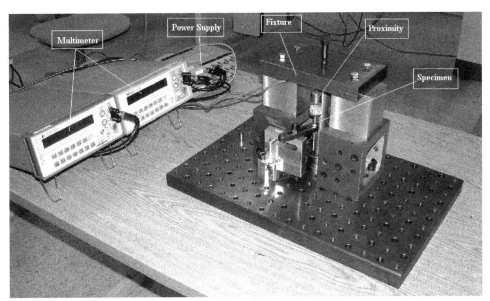

Fig. 9. Measurement Setup

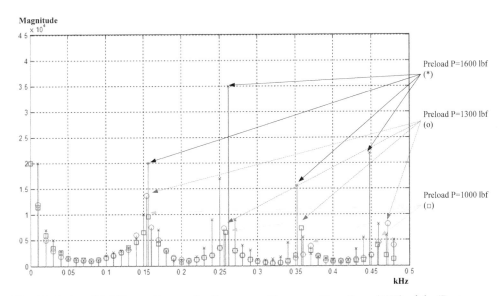

Fig. 10. A simplifed illustration on the Frequency Response Functions (FRF) of the Test System

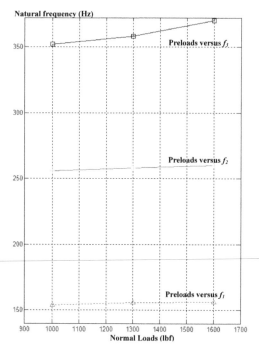

Fig. 11. The relationships between the natural frequencies and preloads

Once the natural frequency is obtained from the test, contact stiffness can be estimated by calculating $\lambda_i l$, β, and k_n. In order to verify the results, these calculations were compared with the previous static measurement results of contact stiffness. Under the same experimental condition, i.e., the same experimental device and preloads, contact stiffness is obtained and used in the calculation of natural frequencies, and then compared with the results of dynamic tests, as shown in Fig.12.

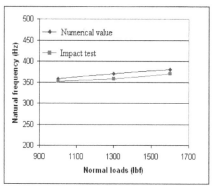
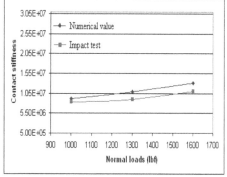

(a) Natural Frequency vs. Preloads   (b) Normal Contact Stiffness vs. Preloads

Fig. 12. Comparisons between Experimental Result of Dynamic Test and Numerical Value Based on Static Test

It can be seen that the results from the dynamic tests are consistent with the numerical calculation results based on the static test results. When the results of the dynamic test are consistent with the static test results, the dynamic test method can be used in identifying tangential contact stiffness, for which the static tests are too difficult to conduct.

4.3 Theoretical formulation of tangential contact stiffness

Two fixture components are in contact at a certain number of asperities due to the inherent roughness of the surface. When they are subjected to tangential forces, the components are mutually constrained through frictional contacts. A friction model based on the Coulomb friction theory is shown in Fig.13. The tangential contact stiffness results from the elasticity of asperities of the contact surfaces, and the total resulting stiffness of these contact surfaces depends on their statistical topographical parameters.

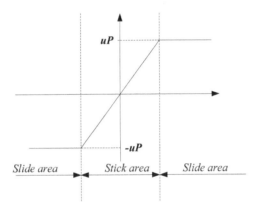

Fig. 13. A Friction Model

Consider that body I is brought into contact with the flat surface of the support under a uniform preload, P, and is subjected to an small excitation, F, as shown in Figure 14. It is assumed that the tangential contact stiffness will change as the preload increases. The friction at each contact point is governed by Coulomb's law. When force is applied in the tangential direction, the asperities in body I will also deform until the shear stress between the asperities exceeds the limit, then the contact surface will slide each other. The friction model of body I in contact is shown in the Fig.15. The friction force is given by Eq. 32.

$$f = \begin{cases} = K_t u & |u| \le \mu P / K_t \\ \mu P & otherwise \end{cases} \tag{32}$$

The idea of the identification of the tangential contact stiffness is to compare the two sets of system natural frequencies: one set is identified from the measured impulse response in tangential direction under different preloads, and the other set is calculated from the FEA model of the system. Based on the numerical simulation, a relationship between tangential contact stiffness and the natural frequencies can be established. If the natural frequencies are measured in the experiments under different preloads, the contact stiffness can be calculated from the relationship obtained by the numerical simulation.

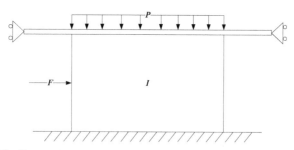

Fig. 14. Body *I* on the Support

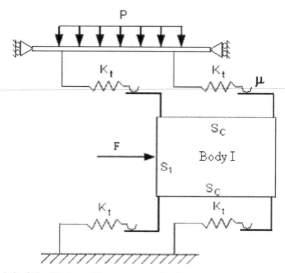

Fig. 15. Friction Model of the Fixture Components in Contact

In order to do the numerical simulation, the effect of the contact force needs to be included into FEA model of the system. The additional contact stiffness matrix will be introduced in the general FEA model. The derivation of contact stiffness matrix is briefly given as follows.

Consider an elastic body I in Fig.15, the kinetic, strain, and potential energies of the system can be expressed respectively as:

$$K = \int_V \frac{\rho}{2} \left(\frac{\partial \{u\}}{\partial t} \right)^T \left(\frac{\partial \{u\}}{\partial t} \right) dV \tag{33}$$

$$U = \int_V \frac{1}{2} \{\sigma\}^T \{\varepsilon\} dV \tag{34}$$

$$W = - \left(\int_{S_1} \{\hat{F}\}^T \{u\} dS + \int_{S_C} \{\hat{R}_c\}^T \{u\} dS \right) \tag{35}$$

where K is the kinetic energy; $\{u\}$ is the displacement vector; V is the volume of the elastic body I; ρ is the mass density of the material; U is the strain energy; $\{\varepsilon\}$ and $\{\sigma\}$; are the strain and stress components, respectively. W is the potential energy of external forces; $\{\hat{F}\}$ is the external surface force vector specified on the boundary S_I; $\{\hat{R}_c\}$ is the contact force vector on the contact surface S_c. Note that $S_c \cap S_1 = \{\Phi\}$; The body force is ignored. Using the above energy expressions the total potential energy of the system is

$$\Pi = K - (W + U) \tag{36}$$

Based on the well-known Hamilton's principle, a discretized FEA formulation for a typical element can be expressed as

$$\left[M^e\right]\{\ddot{d}(t)\} + \left\{\left[K^e\right] + \left[\tilde{K}_C\right]\right\}\{d(t)\} - \{F^e\} = 0 \tag{37}$$

To obtain the matrix form, the displacement field of a typical element $\{u\}$, which is a function of both space and time, can be written as:

$$\{u\} = \left[N(x)\right]\{d(t)\} \quad \{\dot{u}\} = \left[N(x)\right]\{\dot{d}(t)\} \quad \{\ddot{u}\} = \left[N(x)\right]\{\ddot{d}(t)\} \tag{38}$$

where $[N(x)]$ is a vector of the space function; and $\{d(t)\}$ is the nodal response vector. Using the interpolation relationship the element,

Mass matrix is given by

$$\left[M^e\right] = \int_{V_e} \rho \left[N(x)\right]^T \left[N(x)\right] dV \tag{39}$$

And the element stiffness matrix is

$$\left[K^e\right] = \int_{V_e} [B]^T [D][B] dV_e \tag{40}$$

where $[B]$ is the geometry matrix.

Comparing to the standard FEA formulation an additional term of $\tilde{K}_C$, referred as the contact stiffness matrix in included in Eq. (37). The term stems from the work done by the contact force on the contact surface. A brief derivation is presented as follows.

The work done by the contact force on the contact surface can be written as

$$E_{ce} = \int_{S_{Ce}} \{u\}^T \{\hat{R}_{ce}\} dS \tag{41}$$

Using the contact element, the contact force can be expressed as

$$\{\hat{R}_{ce}\} = [D_c]\{u\} \tag{42}$$

Substituting Eqs. (38) and (42) into (41) yields

$$E_{ce} = \{d(t)\}^T \int_{S_{Ce}} [N(x)]^T [D_c][N(x)]dS\{d(t)\} \qquad (43)$$

Therefore, the contact stiffness matrix can thus be defined as

$$[\tilde{K}_C] = \int_{S_{Ce}} [N(x)]^T [D_c][N(x)]dS \qquad (44)$$

where $[D_c]$ is the contact property matrix. In the section, the displacements of contact element in the normal direction are assumed to keep stick. Therefore, the normal contact stiffness becomes infinity. The tangential contact stiffness is considered.

The derived contact stiffness matrix should be added to the general FEA model for the fixture stiffness analysis to take into account the effects of the contact force. Followed the standard procedure of the eigenvalue problem, the system natural frequencies can be obtained using the FEA method to establish the relationship between the tangential contact stiffness and natural frequencies. For example, a specimen that has the dimensions 5×3×0.75 in was used to measure dynamic characteristics. Fig.16 shows the FEA model of the specimen. Contact elements were modeled as separate springs on the top and bottom surfaces of the specimen. There are two nodes for each contact element. One node is on the contact surface of the specimen. The other node is constrained at all degrees of freedom. The impulse force was applied at the side of the specimen. The response was obtained at point M, at the other side of the specimen.

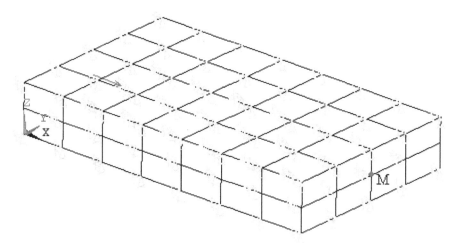

Fig. 16. Finite Element Model of Specimen

Fig.17 shows the relationships between tangential contact stiffness and natural frequencies of the first two vibration modes. The results are obtained through numerical simulation. From experiments, the frequency response is measured under the different preloads. The contact stiffness can be determined based on the relationships shown in Fig.17.

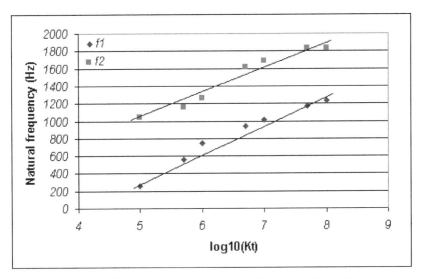

Fig. 17. The Relationship of Tangential Stiffness vs. the First Two Natural Frequencies

5. Conclusions

Forces in a workpiece-fixture system have a crucial impact on the deformation and accuracy of the system. In this chapter, an FEA model of fixture unit stiffness is proposed. A contact model between fixture components are utilized for solving the contact problem encountered in the study of fixture unit stiffness. By several simple experiments and comparison with the corresponding analytical solution and experimental results in the literature, this methodology is validated. This analytic approach also can be extended in the research of complex fixture system with multiple units and components, which will lead to a new progress in the design and verification of fixture-workpiece system study.

6. References

Aliabadi, Brebbia C.A., 1993, *Computational methods in contact mechanics*, Computational Mechanics Publications.

Anthony C. Fischer-Cripps, *Introduction to contact mechanics*, Springer-Verlag Inc, New York. 2000.

Bathe K.J., 1996, *Finite Element Procedures*, Prentic-Hall, Inc.

Beards C.F., 1983, The Damping of Structural Vibration by Controlled Inter-facial Slip in Joints, *Journal of Vibration, Acoustics, Stress, and Reliability in Design*, 105, pp. 369-373

Fang B.; R.E. DeVor; S.G. Kapoor, 2002, Influence of friction damping on workpiece-fixture system dynamics and machining stability, *Journal of Manufacturing Science and Engineering*, 124, pp. 226-233

Hurtado J. H.; S. N. Meltke, 2002, Modeling and Analysis of the Effect of Fixture-Workpiece Conformability on Static Stability, *Journal of Manufacturing Science and Engineering*, 124, pp. 234-241

Lee, J. D.; Haynes L. S., 1987, Finite element analysis of flexible fixturing systems, *Journal of Engineering for Industry*, 109, pp.134-139

Liao, Y. G.; Hu S.J., 2001, An integrated model of a fixture-workpiece system for surface quality prediction, *International Journal of Advanced Manufacturing Technology*, 17, pp. 810-818

Mazurkiewicz M; W. Ostachowicz, 1983, Theory of finite element method for elastic contact problems of solid bodies, *Computer Structure*, Vol. 17, pp.51-59

Rong, Y.; Zhu Y., 1999, *Computer-Aided Fixture Design*, New York, NY, Dekker.

Trappey A. J. C.; Su C. S., J. L. Hou, 1995, Computer-aided fixture analysis using finite element analysis and mathematical optimization modeling, ASME, Manufacturing Engineering Division, MED, 2-1, pp.777-787

Zhu, Y.; S. Zhang, Y. Rong, 1993, Experimental study on fixturing stiffness of T-slot based modular fixtures, *NAMRI Transactions XXI*, NAMRC, Stillwater, OK, USA, pp. 231-235

Zheng Y., 2005, *Finite Element Analysis for Fixture Stiffness*, PhD Desertation, Worcester Polytechnic Institute, MA, USA.

Zheng Y.; Hou Z.; Rong Y., 2008, The study of fixture stiffness part I: a finite element analysis for stiffness of fixture units, *International Journal of Advanced Manufacturing Technology*, 36(9-10), pp.865-876

Zheng Y., Hou Z., Rong Y., 2008, The study of fixture stiffness -Part II: contact stiffness identification between fixture components, *International Journal of Advanced Manufacturing Technology*, 38(1-2), pp.19-31

2

Identification of Thermal Conductivity of Modern Materials Using the Finite Element Method and Nelder-Mead's Optimization Algorithm

Maria Nienartowicz and Tomasz Strek
Poznan University of Technology,
Institute of Applied Mechanics,
Poland

1. Introduction

Due to the fact that modern materials are widely used in aerospace industry, automotive industry, medicine and many others it is very important to find out a way to meet all its needful parameters. There is a lot of composites, which arise from a combination of at least two different components on the macroscopic level and whose parameters are unknown. Therefore, it is difficult to get to know their possibilities and functionality. A kind of composite materials are Functionally Graded Materials (FGM). They are characterized by the fact that its composition and structure gradually change over the volume, which follows from changes in properties of material (Miyamoto et al., 1999). There are many works on this topic.

Microstructure and thermal stress relaxation of ZrO2-Ni, which is an example of FGM, by hot-pressing was studied in (Jingchuan et al., 1996). The researches consisted of scanning, transmission electron microscopy and X-ray diffractometry shown that the chemical composition and microstructure of ZrO2-Ni FGM is distributed gradiently in stepwise way. The preliminary analyze of thermal stress distribution by means of Finite Element Method was also presented there. In other article the identification of the nonlinear thermal-conductivity coefficient by gradient method was shown (Borukhov, 2005). Author puts his attention to the gradient methods of solution of the inverse heat-conduction problem of designation of the nonlinear coefficient $\lambda(T)$, without preliminary finite-dimensional approximation. Review of the principal developments in Functionally Graded Materials, with an emphasis on the recent work published since 2000, are presented in (Birman&Byrd, 2007). A various areas relevant to the different aspects of theory and applications of FGM are submitted there.

As it was mentioned above the difficulty of working with composites is that many parameters cannot be directly determined. An example would be the relationship between composite properties and temperature variation. The study of these relationships can be done experimentally, which is costly because it requires proper equipment and sample preparation.

The effect of multi fibres filler in composite on thermal conductivity was examined in paper (Jopek&Strek, 2011). Three types of optimization were performed in terms of effective thermal conductivity: minimization, maximization and determination of arrangement which gives expected value of effective thermal conductivity. Hybrid method combining optimization with genetic algorithm and differential equation solver by finite element method was used to find optimal arrangement of fibers position in composite matrix.

In this paper an attempt to perform simulation for comparison numerical results with experimental results was made. To carry out the simulation, the environment of the program COMSOL Multiphysics was used (Comsol, 2007). The code was written in COMSOL Script and combines Finite Element Method (Zienkiewicz&Taylor, 2000) with optimization Nelder-Mead algorithm (Nelder&Mead, 1965). This algorithm belongs to non-gradient optimization methods of function of many variables (Weise, 2009). The computations are performed using the simplex, so that in each iteration value at several specific points is calculated. In this way we find the minimum of the function. It found its application in many studies and wide range of problems.

This method was used for engineering optimization as a Globalized Bounded Nelder-Mead algorithm (GBNM) (Luersen&Le Riche, 2004). In that article a global approach to real optimization was shown by using restart procedure. The Globalized Bounded Nelder-Mead can be applied to discontinuous, non-convex functions. To speed up a global search an improved restart procedure was found. An example of this was shown in article where improved restart procedure was used for optimization of composite bracket (Hossein Ghiasi et al., 2007). This approach was changed by using a one-dimensional adaptive probability function and including nonlinear constraints. Thereby the Improved Globalized Nelder-Mead Method became more efficient than evolutionary algorithm, as results confirmed. Some attempts of benchmarking the Nelder-Mead downhill simplex algorithm with many local restarts appeared in (Hansen, 2009). This method was also applied for Multiple Global Minima (Stefanescu, 2007). In that article author proves that Nelder-Mead heuristic procedure can detect successfully multiple global minima.

Simplex method, mentioned above, can be also used as combined with genetic algorithm. In this way the genetic algorithm is used to find a global optimum area and then the Nelder-Mead algorithm is used for a local optimization (Durand&Alliot, 1999). A hybrid genetic and Nelder-Mead algorithm (HGNMA) was also used for decoupling of Multiple Input Multiple Output (MIMO) system with application on two coupled distillation columns process (Lasheen et al., 2009). In that article a technique that uses relative gain array to choose proper pairing and HGNMA to find optimal elements' values of the steady state decoupling compensator unit was proposed. That minimizes internal couplings of MIMO systems. Similar hybrid was presented for optimization in the variational methods of Boundary Value Problems (Mastorakis, 2009). Author presents a way of solution of p-Laplacian equation. Next, it is discussed with other methods for the solution. Using the Nelder-Mead's method also problems of identification of material parameters can be solved, as it is shown in article where investigation of processes in a rock mass is described (Blaheta et al., 2010).

Simulation, which is carried out in this chapter, for the heat transfer in considered domains with the boundary conditions in the form of a heat flux on both ends or temperature allows to determine parameters such as a thermal conductivity, a heat exchange surface area at the boundary or an outside temperature around both ends of the area. Modifying the program code simulations which optimize the temperature, in case when the thermal conductivity is dependent on the nonlinear function, can be also performed. The Nelder-Mead's optimization algorithm combined with Finite Element Method calculations are performed in COMSOL Script environment. Considered modern composite consists of multiple materials with different properties in different sections.

2. Heat transfer

The heat transfer can be defined as a movement of energy which is caused of temperature difference. It can be provided by the three mechanisms. First of them is a conduction, which can be described as diffusion, which is held in a stationary medium and occurs because of temperature gradient. The mentioned medium can be in form of solid or fluid. Next is convective, which appears as a result of fluid motion. The last one is radiation, which follows from electromagnetic waves between two surfaces, on which different temperatures are. Additionally those surfaces must comply with a condition that the first surface is visible to an infinitesimally small observer on the second surface.

The heat transfer by conduction can be defined by the heat equation

$$\rho \cdot C_p \cdot \frac{\partial T}{\partial t} - \nabla \circ (k \cdot \nabla T) = Q \,, \tag{1}$$

where: T – is the temperature, ρ – is the density, C_p - is the heat capacity at constant pressure, k – is the thermal conductivity and Q – is a heat source or heat sink. Taking into consideration a steady-state model the temperature does not change with time.

The thermal conductivity describes a relationship between the heat flux vector **q** and the gradient of temperature ∇T (Bejan&Kraus, 2003), so it takes a form of

$$q = -k \cdot \nabla T \,, \tag{2}$$

The heat flux, mentioned above, is a kind of boundary condition, which can be described as

$$n \circ (k \cdot \nabla T) = q_0 + h \cdot (T_{inf} - T) + C_{const} \cdot \left(T_{amp}^4 - T^4\right) \,, \tag{3}$$

where: q_0 - is the input heat flux, $h \cdot (T_{inf} - T)$ - is used for convective heat transfer, where h is the heat transfer coefficient and T_{inf} is the ambient bulk temperature, $C_{const} \cdot \left(T_{amp}^4 - T^4\right)$ - is used for radiation heat transfer, where T_{amp} is the temperature of surrounding radiation environment and C_{const} is a product of surface emissivity and the Stefan-Boltzmann constant.

3. The Nelder-Mead algorithm

The Nelder-Mead algorithm is a method that does not require to determine a derivative of an objective function. This function is determined in few specific points, different in each iteration. The first simplex algorithm was defined by Spendley in 1962. In 1965 Nelder and Mead improved it and turned the simplex search into an optimization algorithm by adding options like: reflection, expansion, contraction and shrinking. Thanks those operations, which speed up the process of optimization, the simplex deforms in way that was suggested by Nelder and Mead to adapt better to the objective functions (Nelder & Mead, 1965).

An n-dimensional simplex with n+1 vertices p0,p1,p2,...,pn is the smallest convex set which contains these points, where set {pj - p0 : 1 ≤ j ≤ n} must consist of linearly independent vector. In the two-dimensional space the simplex can be created from any triangle and in three-dimensional space from any tetrahedra.

In this method selected initial simplex is modified by means of elementary geometric operations called: reflection, expansion, contraction and shrinking. As a result of each of them the vertex, where value of the objective function takes the highest value (the "worst" vertex), is replaced by another – "better". In this way the simplex is coming more and more to local minimum of examined function.

Finding the minimum of the objective function must be preceded by an analysis, where as a result the vertices, in which the objective function takes the smallest and the highest value, are marked in the following way (see Fig. 1a):

- pmin – the vertex where the objective function takes the smallest value:

$$f(pmin) \leq f(pi) \text{ for any } 0 \leq i \leq n \tag{4}$$

- pmax – the vertex where the objective function takes the highest value:

$$f(pmax) \geq f(pi) \text{ for any } 0 \leq i \leq n \tag{5}$$

- $\bar{p}$ – centroid of the points (the vertex pmax is excluded) see:

$$\bar{p} = \frac{1}{n} \cdot \left(\Sigma_{i \neq max} pi \right) \tag{6}$$

After determining points pmin, $\bar{p}$, pmax, a procedure of minimization of the objective function can begin. In each iteration the following stages can be specified: reflection, expansion, contraction and shrinking, which are described below (Weise, 2009).

Reflection – is based on determining a point which is symmetrical image of point pmax relative to point $\bar{p}$. New point is marked as podb (see Fig. 1b) and its coordinates are designed by formula:

$$podb = \bar{p} + \alpha \cdot \left(\bar{p} - pmax \right) \tag{7}$$

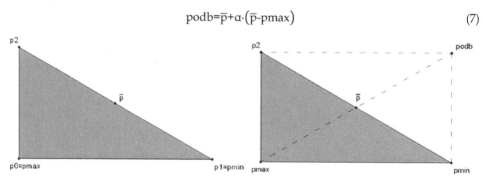

Fig. 1. (a) Initial simplex with vertices p0, p1, p2 where (f(p0)>f(p2)>f(p1)), (b) Simplex: reflection stage

Identification of Thermal Conductivity of Modern Materials Using the Finite Element Method and Nelder-Mead's
Optimization Algorithm

31

Value of reflection coefficient α is in the range of $\alpha \in (0,1]$, but usually it is assumed that $\alpha = 1$.

After reflection stage, depending on value of the objective function in reflection point f(podb), we consider few excluded cases (8), (9) and (10), which determine further investigation in given iterations:

$$f(pmin) \leq f(podb) < f(pmax) , \tag{8}$$

$$f(podb) < f(pmin) , \tag{9}$$

$$f(max) \leq f(podb) , \tag{10}$$

If in calculated point podb the objective function takes value (8) then the reflection is accepted. The new simplex is designed by replacing the vertex pmax with podb. Next indexes min, max and location of point $\bar{p}$ are updated and if a stop condition, which is described later, is not fulfilled a new iteration begins with new reflection.

Expansion – Assume that reflection inequality (9) was fulfilled, which means that a vertex which was found in reflection stage is better point than pmin (it is closer to minimum of objective function f).

It suggests that next steps of finding the minimum should follow in this direction. Because of this the reflection is not accepted and the calculations are carried out by the expansion (see Fig. 2). A new point is calculated and marked as pe:

$$pe=\bar{p}+\gamma\cdot(podb-\bar{p}), \tag{11}$$

where $\gamma>1$ is an expansion coefficient (usually $\gamma = 2$). Next, a value of the objective function in new point is calculated f(pe), and:

- if f(podb) < f(pmin) then the expansion is successful, new simplex is designed by replacing pmax with pe (new simplex is designed by vertices pe, p2, pmin – Fig. 2a); then indexes min and max and a location of point $\bar{p}$ are updated and after checking the stop condition next iteration begins;
- else when f(pe) ≥ f(podb), pmax is replaced by podb (new simplex is designed by vertices podb, p2, pmin – Fig. 2b) and it follows as previous (indexes are updated, stop condition is checked and next iteration begins)

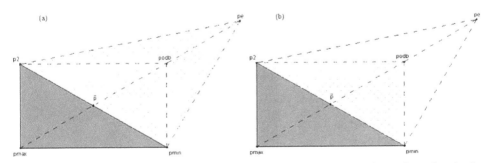

Fig. 2. Expansion stages: successful (a), unsuccessful (b) (new designed simplex is hatched)

Contraction. Reflection cannot be accepted also in case when $f(odb) \geq f(pmax)$, see (10). In this situation occurs contraction of a simplex, whose new vertex is counted according to the formula:

$$pz = \bar{p} + \beta \cdot (pmax - \bar{p}) \qquad (12)$$

where a coefficient of contraction β takes a value $\beta \in (0,1)$, usually $\beta=0.5$ (see Fig. 3a). If point pz leads to improvement, which means $f(pz) < f(pmax)$, then point pmax is replaced by point pz and a new simplex is created (designated by pz, p2, pmin). Next indexes are updated, stop condition is checked and next iteration begins.

Shrinking. This stage takes place when after contraction an inequality (13) is fulfilled:

$$f(pz) \geq f(pmax) \qquad (13)$$

In this situation point pmin remains unchanged, and the whole simplex is shrinking according formula (14):

$$pi \leftarrow \delta \cdot (pi + pmin), \ i=0, 1, ..., n, \ i \neq min \ (14) \qquad (14)$$

where $\delta \in (0,1)$ is a shrinking coefficient and usually $\delta = 0,5$ (see Fig. 3b). A simplex which is build of a new obtained points p0, ..., pn is used in next iteration (if the stop condition is not fulfilled).

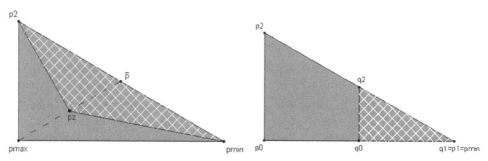

Fig. 3. (a) Contraction stage (if contraction is successful the hatched simplex is chosen), (b) Shrinking stage (new designed simplex is hatched)

In this papers two stop conditions were used. The first when an absolute value of difference between $f(pmin)$ and $f(pmax)$ is smaller than accuracy solution

$$abs\left(f(pmin)-f(pmax)\right) < \varepsilon \qquad (15)$$

and the second when a number of iterations is bigger than maximum number of iterations

$$step > maxstep. \qquad (16)$$

Algorithm of Nelder-Mead method

Input data:

Initial simplex with vertices: p0, p1, ..., pn,

Identification of Thermal Conductivity of Modern Materials Using the Finite Element Method and Nelder-Mead's
Optimization Algorithm

33

Coefficients:

α – reflection,

β – contraction,

γ – expansion,

δ – shrinking,

ε - accuracy of solution,

maxstep – maximum number of iterations.

1. Repeat
2. count a value of function in vertices of simplex: p0, p1, ..., pn
3. find pmin, pmax (min≠max)
4. $\bar{p}=\frac{1}{n}\cdot(\sum_{i\neq max} pi)$,
5. podb=$\bar{p}$+α·$(\bar{p}$-pmax$)$
6. if f(podb) < f(pmin) then
7.   pe=$\bar{p}$+γ·$(podb-\bar{p})$
8.   if f(pe)<f(podb) then
9.   pmax=pe   ▶ expansion
10.   else
11.   pmax=podb   ▶ reflection
12.   end if
13. else
14.   if f(pmin)≤f(podb)<f(pmax) then
15.   pmax=podb
16.   else
17.   pz=$\bar{p}$+β·$(pmax-\bar{p})$
18.   if f(pz)≥f(pmax) then
19.   for i=0 to n do
20.   if i≠min then
21.   pi=δ·(pi+pmin)   ▶ shrinking
22.   end if
23.   end for
24.   else
25.   pmax=pz   ▶ contraction
26.   end if
27.   end if
28. end if
29. until abs$\left($f(pmin)-f(pmax)$\right)$ < ε or step > maxstep   ▶ stop conditions
30. return x*= pmin   ▶ approximate solution

4. Reconstruction of thermal parameters in 1-D domain

As a first a reconstruction of thermal parameters was carried out. This simulation was made for heat transfer in 1-D space, in a domain which length was 1 m. The boundary condition was the heat flux at both ends of the domain. Basing on the temperature distribution in area $T(x)$ thermal parameters of the issue were designated. Those parameters were: a thermal conductivity (k), a transversal convective heat transfer coefficient ($h1$ and $h2$) and an external temperature around both ends of the domain (*Tinf1* and *Tinf2*). The reconstruction of mentioned factors was performed by the optimization.

4.1 Stage I

In the first stage desired temperature distribution was defined as

$$\overline{T}(x)=18.75 \cdot x+287.5,\qquad(17)$$

and minimized integral have a form of

$$F = \int_0^1 \text{Abs}\left(T(x)-\overline{T}(x)\right) dx = \int_0^1 \text{Abs}\left(T(x)-(18.75 \cdot x+287.5)\right) dx.\qquad(18)$$

The start simplex for particular parameters is presented in Table 1.

Vertices	$\sqrt{k}$	$\sqrt{h1}$	$\sqrt{h2}$	$\sqrt{Tinf1}$	$\sqrt{Tinf2}$
p_1	1	2	3	4	5
p_2	100	2	3	4	5
p_3	1	100	3	4	5
p_4	1	2	100	4	5
p_5	1	2	3	100	5
p_6	1	2	3	4	100

Table 1. Start simplex for the stage I

Numbers, placed in Table 1, are square roots of searched thermal parameters. This assumption guarantees that values will be positive. Results of the calculations are presented below in Table 2 and Table 3. Required accuracy of solution was obtained after 88 steps with $F=0.0105$ for following set of parameters.

Parameter	$\sqrt{k}$	$\sqrt{h1}$	$\sqrt{h2}$	$\sqrt{Tinf1}$	$\sqrt{Tinf2}$
p_{min}	53.1230	16.4196	42.9727	9.5704	-18.2983

Table 2. Values of minimized square roots of pmin

Parameter	k	$h1$	$h2$	$Tinf1$	$Tinf2$
p^2_{min}	2800	269.6035	1800	91.5928	334.8276

Table 3. Values of minimized parameters pmin

The objective function F was minimized with the accuracy of solution $\varepsilon=1e-2$.

Identification of Thermal Conductivity of Modern Materials Using the Finite Element Method and Nelder-Mead's
Optimization Algorithm

35

4.2 Stage II

In the second stage of calculation some restrictions were imposed to the optimal parameters, as it is presented below:

$$80 < k < 120,$$

$$8 < h1 < 12,$$

$$15 < h2 < 25,$$

$$80 < Tinf1 < 120,$$

$$350 < Tinf2 < 45.$$

Thereby, we did not have to minimize the roots of objective function.

The start simplex for particular parameters is presented in Table 4.

k	$h1$	$h2$	$Tinf1$	$Tinf2$
80	8	15	80	350
120	12	25	120	450
82	9	18	90	360
110	11	24	119	420
101	8	17	89	360
110	11	21	111	444

Table 4. Start simplex for the stage II

Numbers placed in Table 4 are not square roots of searched thermal parameters, but there are values which minimize the objective function.

Required accuracy of solution was obtained after 45 steps with $F=0.1801$ for following set of parameters.

Parameter	k	$h1$	$h2$	$Tinf1$	$Tinf2$
p_{min}	101.0347	9.8034	20.0654	101.9420	396.8020

Table 5. Values of minimized parameters pmin

Values which were determined are within established limits.

4.3 Stage III

Next stage of the research was optimization of the thermal parameters of material in which coefficient of thermal conductivity was dependent on spatial variable x, like in the Functionally Graded Materials. In those composites temperature distribution with given boundary conditions are usually nonlinear. In these paper it is assumed that parameter $k(x)$ has polynomial form

$$k(x)=p_0+p_1x+p_2x^2. \tag{19}$$

In this stage the heat transfer equation takes a form of

$$\nabla\circ(k(x)\cdot\nabla T)=0, \tag{20}$$

where $k(x)$ – thermal conductivity depends on spatial variable x.

The following boundary conditions (different temperature on ends) was assumed for calculations

$$T=T_{01}=283 \text{ K}, \tag{21}$$

$$T=T_{02}=483 \text{ K}. \tag{22}$$

Now the vector of parameters have a form of: $p=[p_0, p_1, p_2]$.

There were two tasks solved for one function of $\overline{T}(x)$, one integral and for two different start simplexes.

The desired function $\overline{T}(x)$ has a form of:

$$\overline{T}_1(x)=286.25-170.513\cdot x+2398.87\cdot x^2-3898.57\cdot x^3+2133.67\cdot x^4-265\cdot x^5. \tag{23}$$

Minimized integral was defined as:

$$I_1=\int_0^1 \text{Abs}(T(x)-(286.25-170.513\cdot x+2398.87\cdot x^2-3898.57\cdot x^3+2133.67\cdot x^4-265\cdot x^5))dx \tag{24}$$

4.3.1 Task 1

The calculation began with the start simplex described in Table 6.

Vertices, p_{ij}	p_{i0}	p_{i1}	p_{i2}
p_{1j}	10	-150	100
p_{2j}	30	-50	200
p_{3j}	15	-120	120
p_{4j}	25	-80	180

Table 6. Start simplex for the task 1 of stage III

Results were achieved with solution accuracy of 1e-6.

Required accuracy of solution was obtained after 61 steps with $F=2.9492$ for following set of parameters.

Parameter	p_0	p_1	p_2
p_{min}	21.1025	-106.6785	167.7030

Table 7. Values of minimized parameters pmin

Identification of Thermal Conductivity of Modern Materials Using the Finite Element Method and Nelder-Mead's Optimization Algorithm

37

In Fig. 4 disparity between the expected and the obtained temperature distribution is presented. Distribution of the coefficient of thermal conductivity, for $k(x, p_{min})$ is presented in Fig. 5.

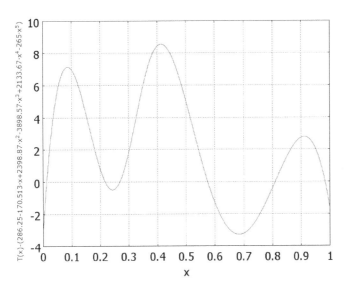

Fig. 4. Disparity between the expected and the obtained temperature distribution (task 1)

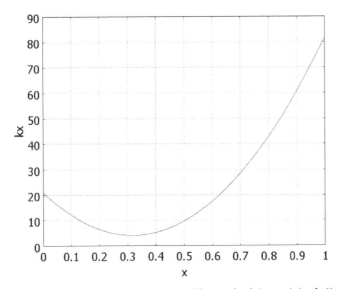

Fig. 5. Distribution of the thermal conductivity coefficient for $k(x, p_{min})$ (task 1)

4.3.2 Task 2

In the second task the start simplex was (Table 8):

Vertices, p_{ij}	p_{i0}	p_{i1}	p_{i2}
p_{1j}	19	-115	130
p_{2j}	16	-95	160
p_{3j}	18	-110	150
p_{4j}	21	-90	170

Table 8. Start simplex for the stage III (task 2)

Results were achieved with solution accuracy of 1e-6.

Required accuracy of solution was obtained after 151 steps with F=2.9336 for the following set of parameters.

Parameter	p_0	p_1	p_2
p_{min}	22.9741	-114.6551	180

Table 9. Values of minimized parameters pmin

In Fig. 6 disparity between the expected and the obtained temperature distribution is presented. Distribution of the coefficient of thermal conductivity, for $k(x, p_{min})$ is presented in Fig. 7.

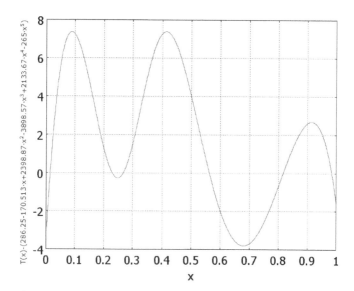

Fig. 6. Disparity between the expected and the obtained temperature distribution (task 2)

Identification of Thermal Conductivity of Modern Materials Using the Finite Element Method and Nelder-Mead's Optimization Algorithm

39

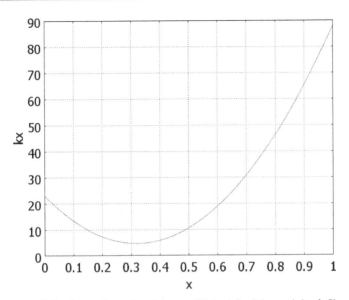

Fig. 7. Distribution of the thermal conductivity coefficient for $k(x,p_{min})$ (task 2)

Despite the fact that the thermal conductivity coefficient seems to look identical in task 1 and 2, there is some difference. In Fig. 8 the disparity in distribution of mentioned thermal conductivity is shown. Concluding, although in task 2 the start simplex was wider (bigger) than in task 1, the solution was found with the same accuracy.

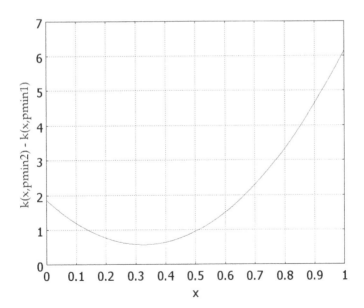

Fig. 8. Disparity between the thermal conductivity coefficient in task 2 and task 1

Summarizing, as it was shown above in this section, it is possible to provide a reconstruction of parameters using hybrid method (FEM with Neleder-Mead). Carrying out the simulation for 1-D domain with length 1 m and defined boundary condition the parameters (the thermal conductivity, the transversal convective heat transfer coefficient and the external temperature around both ends of the domain) can be designated. Values of those parameters can be calculated within some restrictions, which can be specified for material which is examined. It is also possible to designate the value of thermal conductivity parameter of FGM which has a polynomial form.

5. Reconstruction of thermal parameters in 2-D space

In this subsection calculations were made to designate the distribution of the thermal conductivity in 2-D domain. Cylinder with radius r=1m and height z=1m was analysed in 2D axial symmetry model. An axis of symmetry is designated as r=0. In this case it was also assumed that the distribution of the thermal conductivity has a form of polynomial as:

$$k(z)=p_0+p_1 \cdot z+p_2 \cdot z^2+p_3 \cdot z^3. \tag{25}$$

A boundary condition, such that temperature at the top and at the bottom of the cylinder was equal in order that T_{01}=400 K and T_{02}=300 K. For axis r=0 axial symmetry was assumed and on the circumference of the cylinder was assumed a thermal insulation.

In these calculations some restrictions have been imposed. An integral I, which contains sum of three integrals was minimized, as shown below

$$I=I1+5 \cdot (1-I2)+5 \cdot (1-I3), \tag{26}$$

where: I is minimized integral, I1 – is an absolute value of difference between expected and obtained temperature distribution

$$I1 = \int_0^1 \text{Abs}\left(T(z)-\overline{T}(z)\right) dz, \tag{27}$$

I2 – determined part of domain where a relationship such that $k(z)>kmin$, where $kmin$ is a minimum value, is satisfied,

I3 – determined part of domain where a relationship such that $k(z)<kmax$, where $kmax$ is a maximum value, is satisfied.

There were two tasks computed, each for one function of expected temperature. Each task was calculated for two variants of values for $kmin$ and $kmax$ – each of them for three start simplexes, as it is presented in subsection below.

5.1 Task 1 – First function of temperature distribution

In this task the expected temperature distribution took a form of

$$\overline{T}_1(z)=300.481+171.955 \cdot z-72.9167 \cdot z^2, \tag{28}$$

and integral I1 was defined as

$$I1 = \int_0^1 \text{Abs}\left(T(z)-\left(300.481+171.955 \cdot z-72.9167 \cdot z^2\right)\right) dz. \tag{29}$$

Identification of Thermal Conductivity of Modern Materials Using the Finite Element Method and Nelder-Mead's Optimization Algorithm

41

	Vertices, p_{ij}	p_{i0}	p_{i1}	p_{i2}	p_{i3}
Start simplex A (the narrowest)	p_{1j}	20	110	-55	0
	p_{2j}	30	120	-50	1
	p_{3j}	40	130	-45	1
	p_{4j}	50	140	-40	0
	p_{5j}	25	140	-60	1
Start simplex B (wider)	p_{1j}	50	200	-65	5
	p_{2j}	-40	-100	-90	15
	p_{3j}	45	180	5	20
	p_{4j}	80	90	-10	25
	p_{5j}	70	180	-120	30
Start simplex C (the widest)	p_{1j}	100	400	-70	10
	p_{2j}	-50	-200	-130	20
	p_{3j}	50	250	20	100
	p_{4j}	110	20	-150	250
	p_{5j}	100	200	-300	70

Table 10. Task 1 - Different start simplexes taken for calculation

Three different start simplexes were assumed and collected in Table 10. As it was mentioned above there were two variants of calculations. Results and assumptions for them are presented in subsections 5.1.1 and 5.1.2.

5.1.1 Variant 1

For calculations below we defined restrictions as follows: kmin=20 and kmax=120. Which means that we were looking for k(z) distribution in this range: 20<k(z)<120. Numerical results are presented in table below.

Simplex	Steps	p_{i0}	p_{i1}	p_{i2}	p_{i3}	Fmin
A	255	1.98993e+1	2.71751e+1	-3.02331e+1	8.22635e+1	1.54875e-1
B	208	2.11709e+1	1.24838e+1	1.17308e+1	4.99511e+1	3.02988e-1
C	121	1.99063e+1	2.77594e+1	-3.17098e+1	8.33822e+1	1.50866e-1

Table 11. Values of minimized parameters pmin for simplexes A, B, C

For the start simplex A, distribution of the thermal conductivity for minimized value k(pmin) is shown in Fig. 9. Disparity between the expected and the obtained temperature distribution was also examined (see Fig. 10) and it varies between -0.5 and 0.47.

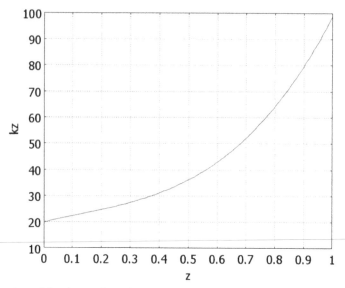

Fig. 9. Distribution of the thermal conductivity (variant 1, start simplex A)

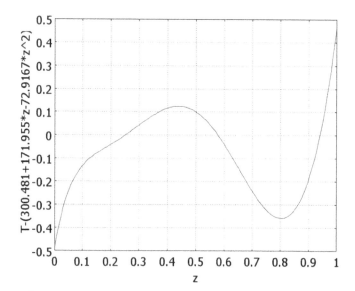

Fig. 10. Disparity between the expected and the obtained temperature distribution
(variant 1, start simplex A)

For the start simplex B, distribution of the thermal conductivity for minimized value
$k(pmin)$ is shown in Fig. 11. Disparity between the expected and the obtained temperature
distribution was also examined (see Fig. 12) and it varies between -0.6 and 0.45.

Identification of Thermal Conductivity of Modern Materials Using the Finite Element Method and Nelder-Mead's
Optimization Algorithm

43

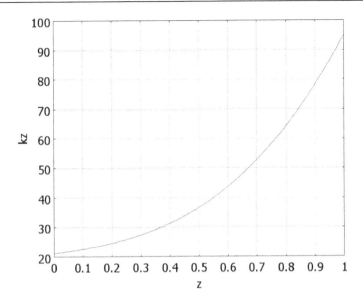

Fig. 11. Distribution of the thermal conductivity (variant 1, start simplex B)

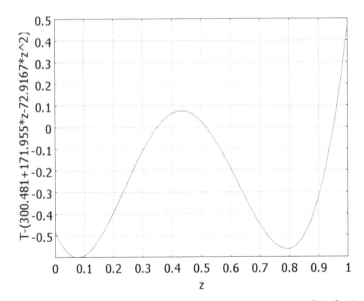

Fig. 12. Disparity between the expected and the obtained temperature distribution
(variant 1, start simplex B)

For the start simplex C, distribution of the thermal conductivity for minimized value
k(pmin) is shown in Fig. 13. Disparity between the expected and the obtained temperature
distribution was also examined (see Fig. 14) and it varies between -0.48 and 0.46.

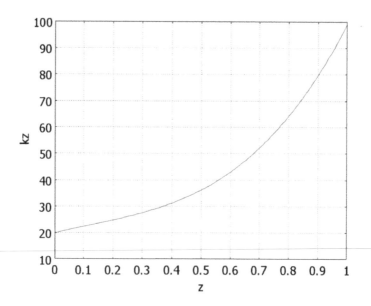

Fig. 13. Distribution of the thermal conductivity (variant 1, start simplex C)

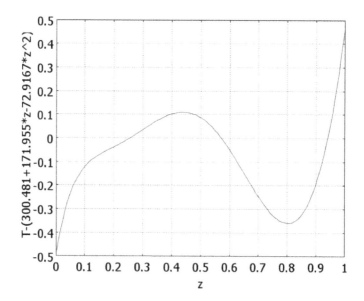

Fig. 14. Disparity between the expected and the obtained temperature distribution (variant 1, start simplex C)

In all cases the results were achieved with solution accuracy of 1e-5. Temperature distribution was similar for all simplexes.

5.1.2 Variant 2

For calculations below we defined restrictions as follows: kmin=10 and kmax=320. Which means that we were looking for k(z) distribution in this range: 10<k(z)<320. Numerical results are presented in table below.

Simplex	Steps	p_{i0}	p_{i1}	p_{i2}	p_{i3}	Fmin
A	91	7.81939e+1	2.99943e+2	-6.54279e+1	8.01293	2.21234
B	156	9.98179	5.00957	9.61216	2.01388e+1	3.02264e-1
C	117	3.98227e+1	8.07309e+1	-1.43192e+2	2.37720e+2	8.36632e-2

Table 12. Values of minimized parameters pmin for simplexes A, B, C

For the start simplex A, distribution of the thermal conductivity for minimized value k(pmin) is shown in Fig. 15. Disparity between the expected and the obtained temperature distribution was also examined (see Fig. 16) and it varies between -3.8 and 2.8.

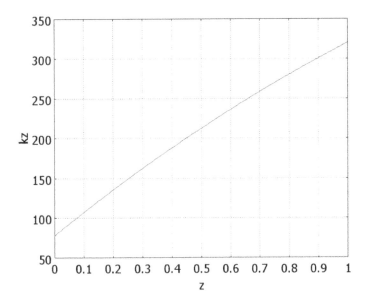

Fig. 15. Variant 2, start simplex A- Distribution of the thermal conductivity

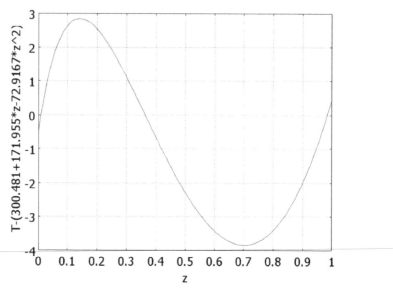

Fig. 16. Variant 2, start simplex A - Disparity between the expected and the obtained temperature distribution

For the start simplex B, distribution of the thermal conductivity for minimized value k(pmin) is shown in Fig. 17. Disparity between the expected and the obtained temperature distribution was also examined (see Fig. 18) and it varies between -0.56 and 0.46.

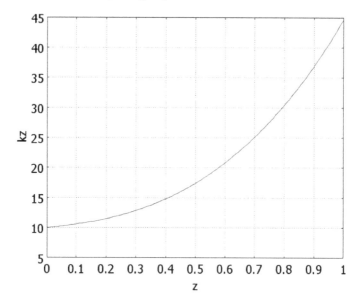

Fig. 17. Variant 2, start simplex B- Distribution of the thermal conductivity

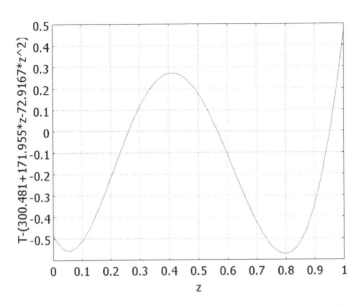

Fig. 18. Variant 2, start simplex B - Disparity between the expected and the obtained temperature distribution

For the start simplex C, distribution of the thermal conductivity for minimized value k(pmin) is shown in Fig. 19. Disparity between the expected and obtained temperature distribution was also examined (see Fig. 20) and it varies between -0.48 and 0.47.

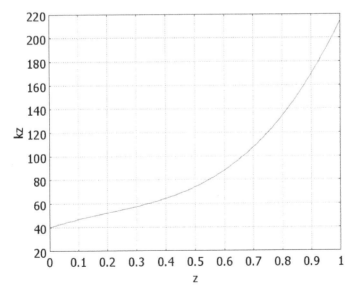

Fig. 19. Variant 2, start simplex C- Distribution of the thermal conductivity

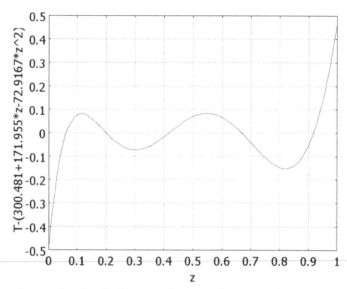

Fig. 20. Variant 2, start simplex C - Disparity between the expected and the obtained temperature distribution

In all cases the results were achieved with solution accuracy of 1e-5. Temperature distribution was similar for all simplexes.

5.2 Task 2 – Second function of temperature distribution

The form of the second expected temperature distribution looks as follows:

$$\overline{T}_1(z)=298.794+149.678\cdot z-47.9503\cdot z^2,\tag{30}$$

and integral I1 takes a form of

$$I1 = \int_0^1 \mathrm{Abs}\left(T(z)-\left(298.794+149.678\cdot z-47.9503\cdot z^2\right)\right)dz .\tag{31}$$

For this task another three different start simplexes were assumed and collected in Table 13.

	Vertices, p_{ij}	p_{i0}	p_{i1}	p_{i2}	p_{i3}
Start simplex A (the narrowest)	p_{1j}	10	150	-80	0
	p_{2j}	15	160	-90	1
	p_{3j}	30	210	-110	1
	p_{4j}	40	220	-120	0
	p_{5j}	50	230	-130	1
Start simplex B (wider)	p_{1j}	100	400	-90	20
	p_{2j}	-60	-300	-230	40
	p_{3j}	90	350	60	200
	p_{4j}	40	220	-120	0
	p_{5j}	150	200	-300	90

Identification of Thermal Conductivity of Modern Materials Using the Finite Element Method and Nelder-Mead's
Optimization Algorithm

49

	Vertices, p_{ij}	p_{i0}	p_{i1}	p_{i2}	p_{i3}
	p_{1j}	150	150	-95	30
	p_{2j}	-80	-350	-290	50
Start simplex C (the widest)	p_{3j}	100	380	70	220
	p_{4j}	50	280	-180	20
	p_{5j}	200	250	-400	100

Table 13. Task 2 - Different start simplexes taken for calculation

For this function also two variants were calculated. Results and assumptions for them are presented in chapters 5.2.1 and 5.2.2.

5.2.1 Variant 1

The restrictions defined for these variants take a form of: kmin=20 and kmax=120. This means that we were looking for k(z) distribution in range like: 20<k(z)<120. Numerical results are presented in table below.

Simplex	Steps	p_{i0}	p_{i1}	p_{i2}	p_{i3}	Fmin
A	138	5.84429e+1	-1.02247e+2	3.68435e+2	-2.21308e+2	6.60791e-1
B	111	3.95147e+1	5.94787e+1	-1.13091e+2	1.3497e+2	3.07911e-1
C	240	4.55508e+1	1.59731e+1	7.83838	5.09419e+1	2.02351e-1

Table 14. Values of minimized parameters pmin for simplexes A, B, C

For the start simplex A, distribution of the thermal conductivity for minimized value k(pmin) is plotted in Fig. 21. Between the expected and the obtained temperature distribution was some disparity which is from -1.48 and 1.2 (see Fig. 22).

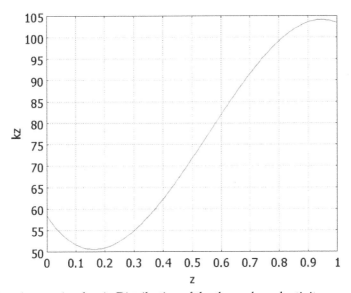

Fig. 21. Variant 1, start simplex A- Distribution of the thermal conductivity

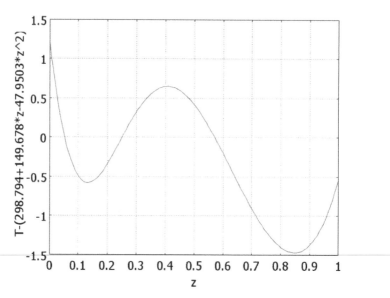

Fig. 22. Variant 1, start simplex A - Disparity between the expected and the obtained temperature distribution

For the start simplex B, distribution of thermal conductivity for minimized value k(pmin) is plotted in Fig. 23. It was some disparity, between the expected and the obtained temperature distribution, and value of it was from -0.45 to 1.21 (see Fig. 24).

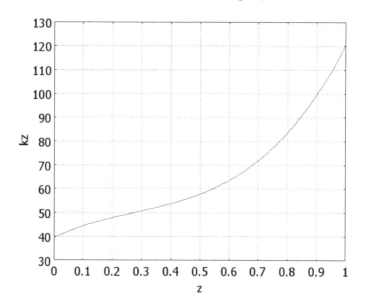

Fig. 23. Variant 1, start simplex B- Distribution of the thermal conductivity

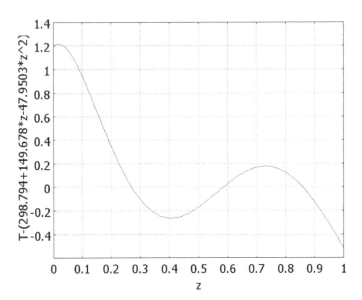

Fig. 24. Variant 1, start simplex B - Disparity between the expected and the obtained temperature distribution

For the start simplex C, distribution of the thermal conductivity for minimized value k(pmin) is plotted in Fig. 25. It was some disparity, between the expected and the obtained temperature distribution, and value of it was from -0.45 to 1.2 (see Fig. 26).

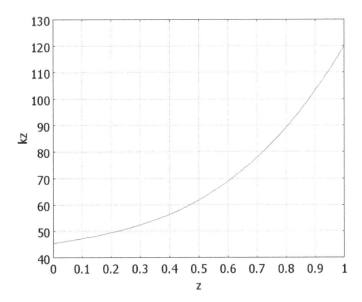

Fig. 25. Variant 1, start simplex C- Distribution of the thermal conductivity

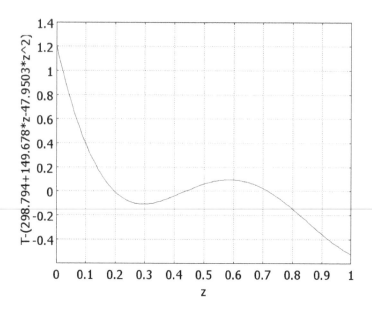

Fig. 26. Variant 1, start simplex C - Disparity between the expected and the obtained temperature distribution

5.2.2 Variant 2

The restrictions defined for these variants take a form of: kmin=10 and kmax=320. This means that we were looking for k(z) distribution in range like: 10<k(z)<320. Numerical results are presented in table below.

Simplex	Steps	p_{i0}	p_{i1}	p_{i2}	p_{i3}	Fmin
A	129	1.34412e+2	-1.60382e+2	6.15741e+2	-3.19053e+2	4.65232e-1
B	146	1.13668e+2	9.88941e+1	-1.36084e+2	2.45597e+2	2.31297e-1
C	119	1.32751e+2	-2.95152e+1	2.27876e+2	-9.65976	2.4153e-1

Table 15. Values of minimized parameters pmin for simplexes A, B, C

In Fig. 27 the distribution of the thermal conductivity for minimized value k(pmin) is plotted for the start simplex A. It was some disparity, between the expected and the obtained temperature distribution, and value of it was from -1 to 1.2 (see Fig. 28).

Identification of Thermal Conductivity of Modern Materials Using the Finite Element Method and Nelder-Mead's Optimization Algorithm

53

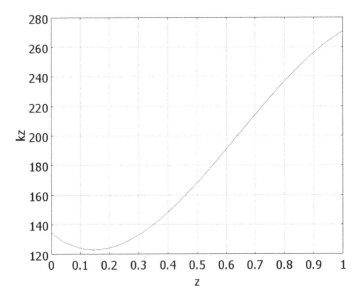

Fig. 27. Variant 2, start simplex A- Distribution of the thermal conductivity

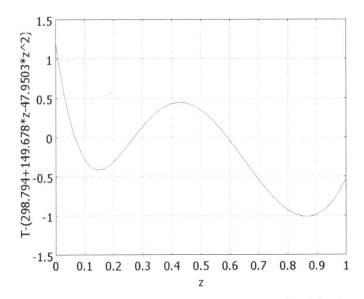

Fig. 28. Variant 2, start simplex A - Disparity between the expected and the obtained temperature distribution

In Fig. 29 distribution of the thermal conductivity for minimized value k(pmin) is plotted for the start simplex B. It was some disparity, between the expected and the obtained temperature distribution, and value of it was from -0.41 to 1.2 (see Fig. 30).

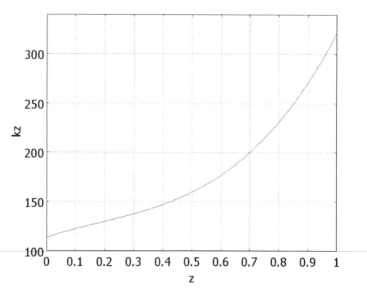

Fig. 29. Variant 2, start simplex B- Distribution of the thermal conductivity

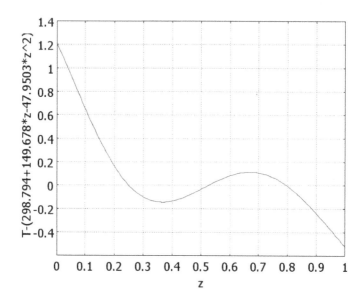

Fig. 30. Variant 2, start simplex B - Disparity between the expected and the obtained temperature distribution

In Fig. 31 the distribution of thermal conductivity for minimized value k(pmin) is plotted for the start simplex C. It was some disparity, between the expected and the obtained temperature distribution, and value of it was from -0.47 to 1.2 (see Fig. 32).

Identification of Thermal Conductivity of Modern Materials Using the Finite Element Method and Nelder-Mead's Optimization Algorithm

55

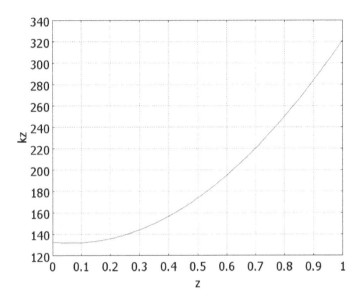

Fig. 31. Variant 2, start simplex C- Distribution of the thermal conductivity

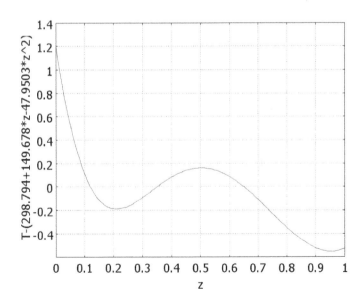

Fig. 32. Variant 2, start simplex C - Disparity between the expected and the obtained temperature distribution

In all cases the results were achieved with solution accuracy of 1e-5. For all simplexes the temperature distribution was similar.

Summarizing, in this chapter some new possibilities of Identification of Thermal Conductivity of Modern Materials using the Finite Element Method and Nelder – Mead's Optimization Algorithm were proposed. Simulating heat transfer in 2-D axial symmetry model, made of the Functionally Graded Material, it is possible to designate its thermal conductivity distribution. This parameter can have the polynomial form. It is also possible to calculate its values within some restrictions. All solutions were found regardless of how far the start simplex was from the solution.

6. Conclusion

Because experimental evaluation of thermal parameters of composites is expensive and time consuming, computational methods have been found to be efficient alternatives for predicting the best parameters of composites. As it was presented in this chapter the Nelder-Mead algorithm connected with the Finite Element Method can be used to optimize many different issues. It has its applicable in problems where it is difficult or impossible to designate the gradient of the objective function. The developed hybrid method can be used for optimization of the heat transfer problems.

In the section 4 of this chapter reconstruction of parameters was provided. Some heat transfer parameters in one-dimensional domain with length 1m, for defined boundary conditions were designated using numerical calculations. The thermal conductivity, the transversal convective heat transfer coefficient and the external temperature around both ends of the domain were calculated within some defined restrictions.

Next, in section 5 possibility of designation of the thermal conductivity was shown. The 2-D axial symmetry model was considered where heat transfer was simulated. The thermal conductivity was in polynomial form. There was also possible to put some restrictions on the searched parameters.

The hybrid method, which was proposed here, can be very helpful in designating any parameters of modern materials like for example Functionally Graded Materials. Proposed method can be also used instead of destructive testing of materials.

7. References

Bejan, A. & Kraus, A. D. (2003). *Heat transfer handbook*, WILEY, ISBN 0-471-39015-1, USA.

Birman, V. & Byrd, L. W. (2007). *Modeling and Analysis of Functionally Graded Materials and Structures*, Vol. 60, pp. 195-216. Available from
http://www.ewp.rpi.edu/hartford/~nelsob/EP/REFERENCES/Papers/
FGM%20Modeling%20and%20Analysis.pdf

Blaheta, R., Kohut, R. & Jakl, O. (2010). *Solution of Identification Problems in Computational Mechanics – Parallel Processing Aspects*, Available from
http://vefir.hi.is/para10/extab/para10-paper-70.pdf.

Identification of Thermal Conductivity of Modern Materials Using the Finite Element Method and Nelder-Mead's
Optimization Algorithm

57

Borukhov, V. T. & Timoshpol'skii, V. T. (2005). *Functional Identification of the Nonlinear Thermal-Conductivity Coefficient by Gradient Methods. I. Conjugate Operators*, Journal of Engineering Physics and Thermophysics, Vol. 78, No. 4, pp. 695-702.

Comsol Multiphysics User's Guide (2007). *Modeling Guide and Model Library*, Documentation Set, Comsol AB.

Durand, N. & Alliot, J. M. (1999). *A Combined Nelder-Mead Simplex and Genetic Algorithm*, Available from
http://recherche.enac.fr/opti/papers/articles/gecco99.pdf

Hossein Ghiasi, M., Pasini, D. & Lessard, L. (2007). *Improved Globalized Nelder-Mead Method for Optimization of a Composite Bracket*, Available from
http://www.iccmcentral.org/Proceedings/ICCM16proceedings/contents/pdf/
MonK/MoKA1-04ge_ ghiasimh224461p.pdf

Jingchuan, Z., Zhongda, Y., Zhonghong, L. & Jian, L. (1996). *Microstructure and thermal stress relaxation of ZrO2-Ni Functionally Graded Material*, Vol. 6, No. 4, pp. 94-99.

Jopek, H. & Strek, T. (2011). *Optimization of the Effective Thermal Conductivity of a Composite*, Convection and Conduction Heat Transfer, Amimul Ahsan (Ed.), InTech, ISBN 978-953-307-582-2. Available from
http://www.intechopen.com/articles/show/title/optimization-of-the-effective-
thermal-conductivity-of-a-composite

Lasheen, A. A., El-Garhy, A. M., Saad, E. M. & Eid, S. M. (2009). *Using hybrid genetic and Nelder-Mead algorithm for decoupling of MIMO system with application on two coupled distillation columns process*, Available from
http://www.naun.org/journals/mcs/mcs-124.pdf

Luersen, M. A. & Le Riche, R. (2004). *Globalized Nelder-Mead method for engineering optimization*, Available from
http://www.sciencedirect.com/science/article/pii/S0045794904002378

Mastorakis, N. E. (2009). *Genetic Algorithms with Nelder-Mead Optimization in the variational methods of Boundary Value Problems*, Vol. 8, No. 3, pp. 107-116, ISSN: 1109-2769

Miyamoto, Y., Kaysser, W. A., Rabin, B. H., Kawasaki, A. & Ford, R. G. (1999). *Functionally Graded Materials: Design, Processing and Applications*, Kluwer Academic Publishers, ISBN 0-412-60760-3, USA

Nansen, N. (2009). *Benchmarking the Nelder-Mead Downhill Simplex Algorithm With Many Local Restarts*, Available from
http://hal.archives-ouvertes.fr/inria-00382104/en

Nelder, J. A. & Mead, R. (1965). *Simplex method for function minimization*, Computer Journal, Vol. 7. No. 4, pp 308-313

Stefanescu, S. (2007). *Applying Nelder Mead's Optimization Algorithm for Multiple Global Minima*, Romanian Journal for Economic Forecasting. Available from
http://www.ipe.ro

Weise, T. (2009). *Global Optimization Algorithms– Theory and Application*, Available from
http://www.it-weise.de/

Zienkiewicz, O.C. & Taylor, R.L. (2000). *The Finite Element Method*, Vol. 1-3: The Basis, Solid Mechanics, Fluid Dynamics (5th ed.), Butterworth-Heinemann, Oxford.

3

Modeling of Residual Stress

Kumaran Kadirgama[1], Rosli Abu Bakar[1],
Mustafizur Rahman[1] and Bashir Mohamad[2]
[1]University Malaysia Pahang,
[2]University Tenaga Nasional,
Malaysia

1. Introduction

In many industries, nickel-base alloys represent an important segment of structural materials. Critical components made of these alloys are relied upon to function satisfactorily in corrosive services. The demand for safe, reliable and cost-effective performance requires that these nickel-base alloys provide the anticipated corrosion resistance. Corrosion-resistant high alloy castings are often the subject of major concern because failures of cast components have led to significant downtime costs and operating problems [1]. Over the years, the nickel- chromium-molybdenum / tungsten alloys have proven to be among the most reliable and cost effective materials for aggressive seawater application and excellent resistance to localized corrosive attack (pitting, crevice corrosion).

Among these alloys, Hestelloy C-types (C, C-4, C-276, and C-22) are used to serve the above mentioned purposes. As these alloys are commonly subject to further machining after casting, it becomes very vital to have an idea about the change in properties imparted to the machined surfaces after such cutting operations as end milling. For this reason, finite element methodology is used in this study to determine the machined surface stress characteristics.

In the past decade, finite element method based on the updated-Lagrangian formulation has been developed to analyze metal cutting processes [1-7]. Several special finite element techniques, such as the element separation [1-7], modeling of worn cutting tool geometry [1, 2, 4, 5, 6], mesh rezoning [3, 5], friction modeling [1-7], etc. have been implemented to improve the accuracy and efficiency of the finite element modeling. Detailed work-material modeling, which includes the coupling of temperature, strain-rate, and strain hardening effects, has been applied to model material deformation [3, 5, 6]. An early analytical model for predicting residual stresses was proposed by Okushima and Kakino [8], in which residual stresses were related to the cutting force and temperature distribution during machining. In another analytical model [9] a connection was made between residual stresses and the hardness of the workpiece. Shih and Yang [10] conducted a combined experimental/computational study of the distribution of residual stresses in a machined workpiece. More recently, Liu and Guo [11] used the finite element method to evaluate residual stresses in a workpiece. They also observed that the magnitude of residual stress reduces when a second cut is made on the cut surface. Liu and Barash [12] measured the residual stress on the workpiece subsurface with consideration of tool flank wear. Their findings indicated that under the condition of a lower cutting speed, the mechanical load had a greater impact on residual stress, while the thermal effect become the major factor effecting residual stress at higher cutting speed. Lee and Shaffer [13] proposed a shear-angle model based on the slip-line field theory, which assumes a rigid-perfectly plastic material behavior and a straight shear plane. Kudo [14] modified the slip-line model by introducing a curved shear plane to account for the controlled contact between the curved chip and straight tool face. Henriksen [15] conducted a series of tests to understand residual stresses in the machined surface of steel and cast iron parts under various cutting conditions. Kono et al. [16] and Tonsoff et al. [17] revealed that residual stresses are dependent on the cutting speed. Matsumoto et al. [18] and Wu and Matsumoto [19] observed that the hardness of the workpiece material has a significant influence on the residual stress field. Konig et al. [20] showed that friction in metal cutting also contributes to the formation of residual stresses.

2. Finite element model

The finite element model is composed of a deformable workpiece and a rigid tool. The tool penetrates through the workpiece at a constant speed and constant feed rate. The model assumes plane-strain condition since generally depth of cut is much greater than feed rate. The finite model used in this study is based on the commercial finite element software. The software, called "Thirdwave AdvantEdge" uses six-noded quadratic triangular elements by default.

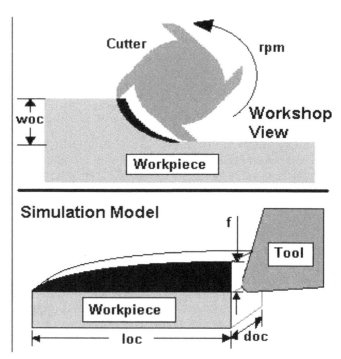

Fig. 1. Thirdwave Advantedge model for milling.

AdvantEdge is an automated program and it is enough to input process parameters to make a two-dimensional simulation of orthogonal cutting operation. The boundary conditions are hidden to the user. Figure 1 shows the Thirdwave AdvantEdge model for milling operation and Figure 2 shows an example of visual simulation of residual stresses induced after milling.

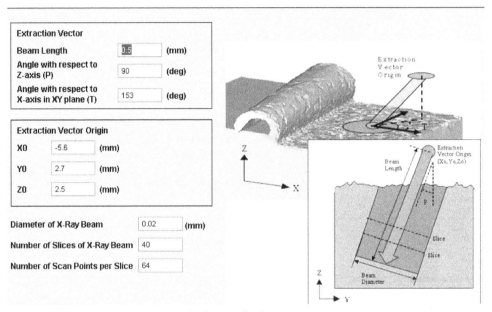

Fig. 2. Thirdwave Advantedge model for residual stress

3. Workpiece and tool material modeling

The workpiece material used for simulation is HASTELLOY C-22HS and the cutting tool is carbide coated with TiALN and 20° rake angle. Every one pass (80mm), the simulation was stopped. AdvantEdge uses an analytical formulation for material modeling. In a typical machining event, in the primary and secondary shear zones very high strain rates are achieved, while the remainder of the workpiece deforms at moderate or low strain rates. In order to account for this, Thirdwave AdvantEdge incorporates a stepwise variation of the rate sensitivity exponent:

$$\overline{\sigma} = \sigma_f\left(\varepsilon^p\right) \cdot \left(1 + \frac{\dot{\varepsilon}^p}{\dot{\varepsilon}_o^p}\right)^{1/m_1}, \ if \ \dot{\varepsilon} \le \dot{\varepsilon}_t^p \tag{1}$$

where $\overline{\sigma}$ is the effective von Mises stress, σ_f is the flow stress, ε^p is the accumulated plastic strain, $\dot{\varepsilon}_o^p$ is a reference plastic strain rate, m_1 is the strain-rate sensitivity exponents, and $\dot{\varepsilon}_t$ is the threshold strain rate which separates the two regimes. In calculations, a local Newton – Raphson iteration is used to compute $\dot{\varepsilon}_o^p$ according to the low – rate equation, and switches to the high rate equation if the result lies above $\dot{\varepsilon}_t$.

σ_f, which is used in Equation (1) is given as:

$$\sigma_f = \sigma_0 \cdot \psi(T) \cdot \left(1 + \frac{\varepsilon^p}{\varepsilon_0^p}\right)^{1/n}$$
(2)

where T is the current temperature, σ_0 is the initial yield stress at the reference temperature T_0, ε_0^p is the reference plastic strain, n is the hardening exponent and $\psi(T)$ is the thermal softening factor. In the present study, it is assumed that the tool is not plastifying. Hence, it is considered as an absolutely rigid body.

4. Results and discussion

Finite Elements simulations were carried out according Table 2. In all simulations, it is made sure that steady- state has been reached and some more data are collected after that time. Therefore, all the results presented in this work were gathered under steady-state condition. From the simulations, variables like stresses, strains, strain rates and temperatures distribution can be obtained. However, all these are very difficult to measure experimentally.

Cutting speed (m/min)	Feedrate (mm/rev)	Axial depth (mm)	Avg. Von Mises stress (Mpa)
140	0.1	2	2100
140	0.2	1	2548
100	0.15	1	1345
100	0.15	2	1548
140	0.15	1.5	2254
100	0.1	1.5	1245
180	0.1	1.5	3987
180	0.15	2	4100
180	0.2	1.5	4488
140	0.2	2	2814
180	0.15	1	4257
140	0.15	1.5	2157
140	0.1	1	1987
100	0.2	1.5	1721
140	0.15	1.5	2347

Table 2: Average value Von mises stress at cutting tool edge

5. Maximum shear stress, stress tensor, Von Mises stress and residual stress

Von Mises stress, σ_V, is used to estimate yield criteria for ductile materials. It is calculated by combining stresses in two or three dimensions, with the result compared to the tensile strength of the material loaded in one dimension. Von Mises stress is also useful for calculating the fatigue strength [21].

Von Mises stress in three dimensions is expressed as[21]:

$$\sigma_V = \sqrt{\frac{(\sigma_1 - \sigma_2)^2 + (\sigma_2 - \sigma_3)^2 + (\sigma_3 - \sigma_1)^2}{2}} \tag{3}$$

where $\sigma_1, \sigma_2, \sigma_3$ are the principal stresses. In the case of plane stress, σ_3 is zero.

Figure 3 shows the Von mises stress for simulation no.9 (Cutting speed 180 m/min, feedrate 0.15 mm and axial depth 2.0 mm) after 80 mm. Most of the tensile σ_V appear at the cutting tool edge. Based on Von Mises criterion, it states that failure occurs when the energy of distortion reaches the same energy for yield/failure in uniaxial tension. Mathematically, this is expressed as [21],

$$\frac{1}{2}\left[(\sigma_1 - \sigma_2)^2 + (\sigma_2 - \sigma_3)^2 + (\sigma_3 - \sigma_1)^2\right] \le \sigma_y^2 \tag{4}$$

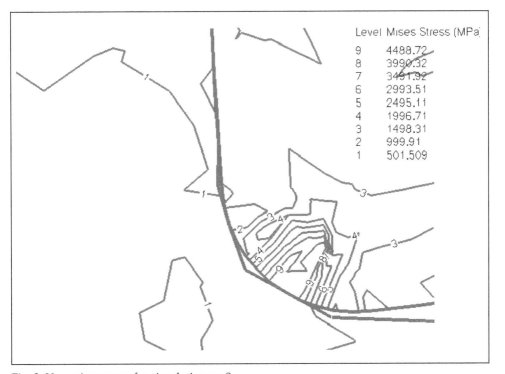

Fig. 3. Von mises stress for simulation no.9.

The yield strength and ultimate tensile strength for the coated carbide cutting tool used in this simulation are 600 MPa and 800 MPa respectively. Then Von Mises stress at at region 9 is 4488 MPa, which is higher than the yield strength and ultimate tensile strength of the coated cutting tool. This strees can cause permanent damage to the cutting tool since this stress is beyond the ultimate tensile strength and yield strength. Cutting speed, feedrate, and axial depth for this simulation is very high and this cause the high stress at the cutting edge, since high cutting speed, feedrate, and axial depth can cause high force in milling [22, 23]. The radial depth for every simulation is 3.5 mm. This factor also contributes to higher stress. At region 1, at the cutting tool and chip contact, the Von mises stress is 501 MPa, where the yield strength and ultimate strength of the workpiece are 359 MPa and 759 MPa. The workpiece start to deform since the stress is above its yield strength.

Figure 4 shows Von Mises stress for simulation no. 3 (Cutting speed 100 m/min, feedrate 0.2 mm/rev, axial depth 1.5 mm). The stress at cutting tool edge (region 5) is 1345 MPa. The Von mises stress is lower compared to that in simulation run no. 9. Even though the stress still higher than yield strength and ultimate tensile strength, but the damage should be not severe compared to that of simulation no.9. At region 3, the stress for the contact point cutting tool and chip is 577 MPa. This value is almost the same as in simulation no.9.

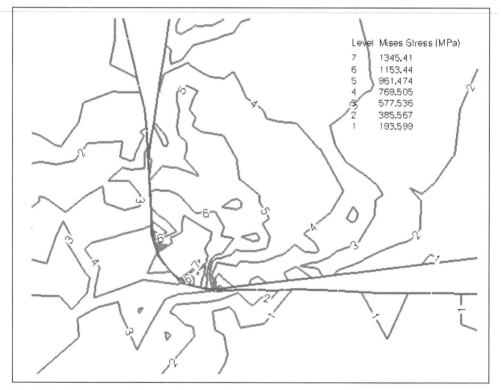

Fig. 4. Von Mises stress for simulation no. 3

From the Von Mises stress distribution as shown in Figure 3 and 4, most of the tensile σ_V locate at the edge of the cutting tool. The stress distribution also show the stress is lower at under the cut surface anFrom the Von Mises stress distribution as shown in Figure 3 and 4, most of the tensile σ_y locate at the edge of the cutting tool. The stress distribution also show the stress is lower at under the cut surface and increases gradually near the cutting edge. High force is needed at the tool edge for workpiece penetration, and this is indirectly increase the stress at the tool edge. This distribution of the stress is same for both cases. The velocity vectors for simulation no.9 as shown in Figure 5 around the tool tip, clearly show the plastic flow of the material araound the cutting edge. The same trend of flow also was observed by M.S.Gadala et. al. [24]. Figure 6 shows the 3D picture for Von misses stress distribution for simulation no.9.

d increases gradually near the cutting edge. High force is needed at the tool edge for workpiece penetration, and this is indirectly increase the stress at the tool edge. This distribution of the stress is same for both cases. The velocity vectors for simulation no.9 as shown in Figure 5 around the tool tip, clearly show the plastic flow of the material araound the cutting edge. The same trend of flow also was observed by M.S.Gadala et. al. [24]. Figure 6 shows the 3D picture for Von misses stress distribution for simulation no.9.

Table 2 show the average value Von mises stress at cutting tool edge for every simulation that already run. This value will be investigate through statistical method to find the relationship between variables (cutting speed, feedrate and axial depth) with response (Von mises Stress).

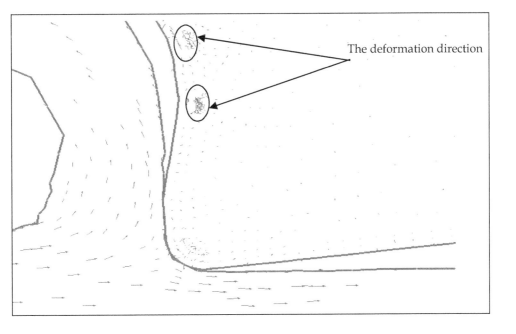

Fig. 5. The velocity vectors for simulation no.9.

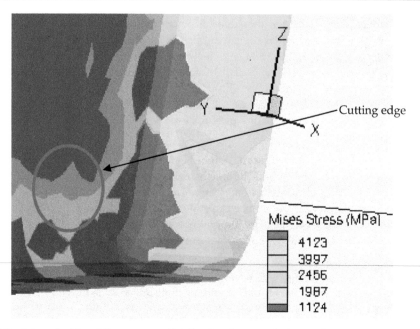

Fig. 6. 3D picture for Von Misses stress distribution for simulation no.9.

6. Conclusion

In the milling operation, cutting speed, feedrate and axial depth play the major role in producing high stresses. The Von Mises stress distribution also show the stress is lower at under the cut surface and increase gradually when come at cutting edge. The highest compressive σ_{xx} appear at the cutting edge. Most of the tensile σ_y appear at the cutting tool edge. The stress distribution also show the stress is lower at under the cut surface and increases gradually near the cutting edge. High force is needed at the tool edge for workpiece penetration, and this is indirectly increase the stress at the tool edge. This distribution of the stress is same for both cases. From the first order model, one can easily notice that the response y (Von Mises stress) is affected significantly by the feed rate followed by axial depth of cut and then by cutting speed. Generally, the increase in feed rate, axial depth and cutting speed will cause Von Mises stress to become larger. The increase in feed rate, axial depth and decrease in cutting speed will cause residual stress to become larger .Response surface method is very useful since with few simulations, a lot of information can be derived such as the relationship between the variables (cutting speed, feedrate and axial depth) with response (Von Mises stress and Residual stress). The combination of numerical analysis and statistical method are very useful to analysis the distribution of stresses in milling.

7. Acknowledgement

The financial support by Malaysian Government through MOSTI and University Tenaga Nasional is grateful acknowledged.

8. References

[1] J. S. Strenkowski and J. T. Carroll, III, A Finite Element Model of Orthogonal Metal Cutting, *ASME 1. Engng Ind.* 107, 345 (1985).

[2] J. S. Strenkowski and G. L. Mitchum, *Proc. North American Manufacturing Conf.*, p. 506 (1987).

[3] A.J.Shih, S. Chandrasekar and H. T. Yang, Fundamental issues in machining, ASME PED-Vol. 43, 11 (1990).

[4] K. Komvopoulos and S. A. Erpenbeek, Finite Element Modelling of orthogonal cutting, *ASME 1. Engng Ind.* 113, 253 (1991).

[5] A.J, Shih and H. T. Yang, Experimental and Finite Element Predictions of Residual Stresses due to Orthogonal Metal Cutting, *Int. I. Numer. Meth. Engng.* 36, 1487 (1993).

[6] A.J. Shih, Finite Element Simulation of Orthogonal Metal Cutting, *ASME 1. Engng Ind.* 117, 84 (1995).

[7] K. Ueda and K. Manabe, Rigid-plastic FEM analysis of three-dimensional deformation fiels in chip formation process *Ann. CIRP* 42, 35 (1993).

[8] K. Okushima, Y. Kakino, The residual stresses produced by metal cutting, Annals of the CIRP 10 (1) (1971) 13–14.

[9] D.W. Wu, Y. Matsumoto, The effect of hardness on residual stresses in orthogonal machining of AISI 4340 steel. Transactions of the ASME, Journal of Engineering for Industry 112 (1990) 245–252.

[10] J. Shih, H.T.Y. Yang, Experimental and finite element predictions of the residual stresses due to orthogonal metal cutting, International Journal for Numerical Methods in Engineering 36 (1993) 1487–1507.

[11] R. Liu, Y.B. Guo, Finite element analysis of the effect of sequential cuts and tool-chip friction on residual stresses in a machined layer, International Journal of Mechanical Sciences 42 (2000) 1069–1086.

[12] C.R. Liu, M.M. Barash, Variables governing patterns of mechanical residual stress in a machined surface. Transactions of the ASME, Journal of Engineering for Industry 104 (1982) 257–264.

[13] E.H. Lee, B.W. Shaffer, The theory of plasticity applied to a problem of machining, Journal of Applied Mechanics 18 (1951) 405–413.

[14] H. Kudo, Some new slip-line solutions for two-dimensional steady state machining, International Journal of Mechanical Science 7 (1965) 43–55.

[15] E.K. Henriksen, Residual stresses in machined surfaces, Transactions ASME Journal of Engineering for Industry 73 (1951) 69–76.

[16] Kono Y, Hara A, Yazu S, Uchida T, Mori Y. Cutting performance of sintered CBN tools, Cutting tool materials. Proceedings of the International Conference, American Society for Metals, Ft. Mitchell, KY, September 15-17, 1980, pp. 218-95.

[17] H.K. Tonshoff, H.G. Wobker, D. Brandt, Tribological aspects of hard turning with ceramic tools, Journal of the Society of Tribologists and Lubrication Engineers 51 (1995) 163–168.

[18] Y. Matsumoto, M.M. Barash, C.R. Liu, Effects of hardness on the surface integrity of AISI 4340 steel. Transactions of the ASME, Journal of Engineering for Industry 108 (1986) 169–175.

[19] D.W. Wu, Y. Matsumoto, The effect of hardness on residual stresses in orthogonal machining of AISI 4340 steel. Transactions of the ASME, Journal of Engineering for Industry 112 (1990) 245–252.

[20] W. Konig, A. Berktold, K.F. Koch, Turning versus grinding—a comparison of surface integrity aspects and attainable accuracy, Annals of the CIRP 42 (1) (1993) 39–43.

[21] J. A. Schey, Introduction to Manufacturing Processes, McGraw Hill,3rd ed, 2000

[22] K.Kadirgama, K.A.Abou-El-Hossein. (2005), 'Force Prediction Model for Milling 618 tool Steel Using Response Surface Methodology', American Journal of Applied Sciences 2(8): 1222-1227, 2005

[23] M.Alauddin,M.A.Mazid, M.A.EL Baradi,M.S.J.Hashmi,"Cutting forces in the end milling of Inconel 718",Journal of Materials Processing Technology",77(1998),pp 153-159

[24] M.Movahhedy, M.S. Gadala, Y. Altintas, "Simulation of the orthogonal metal cutting process using an arbitrary Lagrangian- Eulerian finite –element method", Journal of Materials Processing Technology,103(2000),pp 267-275

4

Reduction of Stresses in Cylindrical Pressure Vessels Using Finite Element Analysis

Farhad Nabhani*, Temilade Ladokun and Vahid Askari
*Teesside University, School of Science and Engineering,
Middlesbrough, TS1 3BA,
UK*

1. Introduction

Due to the differential operating pressure of pressure vessels, they are potentially dangerous and accidents involving pressure vessels can be deadly and poses lethal dangers when vessels contents are flammable/explosive, toxic or reactive. Stress induced operating factors (e.g., process-upset, catalyst regeneration) and stress related defects (e.g., fatigue creep, embrittlement, stress corrosion cracking) accounts for approximately 24.4% of reoccurring catastrophic pressure vessels failures in process industries, many of which has resulted in loss of several lives, properties and in some cases preventive measures of evacuation of residents and community enforced (Sirosh & Niedzwiecki 2008). Pressure vessels store large amounts of energy; the higher the operating pressure - and the bigger the vessel, the more the energy released in the event of a rupture and consequently the higher the extent of damage or disaster or the danger it poses, (American Petroleum Institute 2001). To prevent stress related vessel rupture and catastrophic failure, it is necessary to identify the main factors that contribute extensively to stress development in pressure vessels and how they can be mitigated. This work presents critical design analysis of stress development using 3D CAD models of cylindrical pressure vessels assembly and finite element engineering simulation of various stress and deformation tests at high temperature and pressure.

2. Applications of pressure vessels

Pressure vessels are air-tight containers used mostly in process industry, refinery and petrochemical plant to carry or hold liquid, gases or process fluids. The commonly used types of pressure vessels in the industry are heat exchangers, tanks, towers, boilers, drums, condensers, reactors, columns, air cool exchangers and the usual shape employed in their design are cylinders, cone and spheres as shown in figures below.

Any pressure vessel in-service poses extreme potential danger due to the high pressure and varying operating temperature, hence there should be no complacency about the risks. Unfortunately, pressure vessels accidents happen much more than they should.

---

* Corresponding Author

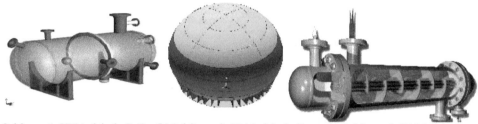

Subfigure 1: 3D Model of a Boiler (b) Subfigure 2: 3D Model of a Drum (c) Subfigure 3: 3D Model of a Heat Exchanger
Source: PM Projects & Services Pvt Ltd.

Fig. 1. Subfigure with three images

2. Rate of pressure vessels accidents

Bulk Transporter (2009) reported that the National Board of Boiler and Pressure Vessel Inspectors in the US recorded the number of accidents involving pressure vessels at an increase of 24% over the course of a year between 1999 and 2000. These statistics include power boilers, steam heating boilers, water heating boilers, and unfired pressure vessels. However, the increased number of accidents was not reflected through to the number of fatalities, as these actually dropped by 33% over this period. By broadening this search, it can be seen that the reporting period of 1992 to 2001 saw a total of 23,338 pressure vessel related accidents which averages at 2,334 accidents per year. Reporting year 2000 saw the highest number of accidents at 2,686 with the lowest at 2,011 in 1998 (National Board of Boiler and Pressure Vessel Inspectors, 2002).

The number of fatalities as a direct result of boiler and pressure vessel accidents has been recorded as 127 over the past 10 years (Air-conditioning, Heating, Refrigeration- The News (2002). During the reported period between 2001 and 2008, the statistics show that the rate of accidents that were directly linked to pressure vessels is not yet on the decline.

3. Causes of pressure vessel failures

The main causes of failure of a pressure vessel are as follows:

- Stress
- Faulty Design
- Operator error or poor maintenance
- Operation above max allowable working pressures
- Change of service condition
- Over temperature
- Safety valve
- Improper installation
- Corrosion
- Cracking
- Welding problems
- Erosion

- Fatigue
- Improper selection of materials or defects
- Low –water condition
- Improper repair of leakage
- Burner failure
- Improper installation Fabrication error
- Over pressurisation
- Failure to inspect frequently enough
- Erosion
- Creep
- Embrittlement
- Unsafe modifications or alteration
- Unknown or under investigation

4. Stresses in pressure vessels

Stress is the internal resistance or counterforce of a material to the distorting effects of an external force or load, which depends on the direction of applied load as well as on the plane it acts. At a given plane, there are both normal and shear stresses (Engineers Edge 2010). However, there are planes within a structural component subjected to mechanical or thermal loads that contain no shear stress. Such planes are principal planes, the directions normal to those planes are principal directions and the stresses are principal stresses. For a general three-dimensional stress state there are always three principal planes along which the principal stresses. (Spence. J & Tooth.A.S. (1994).

Different types of stresses as stated in Chattopadhyay. S. (2004) are as follows:

i.   Pressure stresses
ii.   Thermal stresses
iii. Fatigue stresses
iv.   Local stresses
v.   External stresses
vi.   Compressive stresses
vii. Bending stresses
viii. Normal stress
ix.   Circumferential stresses
x.   Longitudinal stresses
xi.   Radial stresses
xii. Tangential stresses
xiii. Tensile stresses
xiv. Shear stress
xv.   Bending stress
xvi. Principal stress

According to D. R. Moss (2004) stresses are generally categorised as primary, secondary or peak stresses. Primary stresses are stresses due to pressure (internal or external), mechanical loads and wind which can result in the rupture or total collapse of a pressure vessel, they are the most hazardous. Secondary stresses on the other hand are strain-induced stresses, and can be developed at the junction of major components of a pressure vessel (e.g. radial loads on nozzles) because of stresses caused by relenting load or differential thermal expansion. While Peak stresses are the maximum stress concentration point in addition to the primary and secondary stresses present in a region. Peak stresses are only significant in fatigue conditions and are the sources of fatigue cracks, which are applicable to membrane, bending and shear stresses (Rao. K. R. 2002).

When a thin-walled cylinder is subjected to internal pressure, three mutually perpendicular principal stresses will be set up in the cylinder material, namely the circumferential or hoop stress, the radial stress and the *longitudinal* stress, (Sharma.S.C .2010). Provided that the ratio of thickness to inside diameter of the cylinder is less than 1/20, it is reasonably accurate to assume that the hoop and longitudinal stresses are constant across the wall thickness, and that the magnitude of the radial stress set up is so small in comparison with the hoop and longitudinal stresses that it can be neglected. This is obviously an approximation since, in practice, it will vary from zero at the outside surface to a value equal to the internal pressure at the inside surface (Hearn.E.J.1998).

5. Vessel description

The arrangement of a typical pressure vessel is shown in Fig. 2. A typical pressure vessel consists of shell (body of the vessel), closure heads, openings for inspection and instrumentations, and a combination of nozzles for pressure relief or other purpose and supports (Syed U. A. (2009).

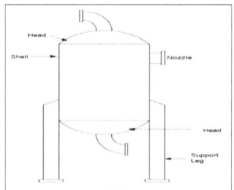
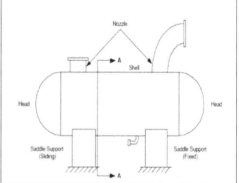

Subfigure 1: Typical horizontal vessel   (b) Subfigure 2: Typical vertical arrangement
Source: Megyesy. E.F (2001)

Fig. 2. Subfigure with two images

6. Finite element model of pressure vessels

In order to proceed with the analysis, three Finite element models were designed and denoted as design case one, two and three. A finite element model consists of boundary conditions, mesh of elements and nodes. Each component of pressure vessel analysed for stress and deformation at the design conditions of 137 MPa and 400°C for all the cases considered. Wind loads and seismic loading were also taken into account. All finite element analyses were run using ANSYS 10.0.

The design parameters taken are as follows:

- Design Code ASME BVPC 2007
- Shell Material SA-723, grade 1, UNS No k23550, class 3
- Design Pressure 137 MPa
- Design Temperature 400 °C
- Tensile Strength of Material 965 MPa
- Material Rating at design temperature 806 MPa
- The shell to be used is fully radiographed, hence E is 100%
- Monobloc shell assumed.

6.1 Model design case one

The pressure vessel model assumed a cylinder with semi ellipsoidal top head and hemispherical bottom head capped. Considering the shell to be monobloc and the design rule that the design pressure P_D should not exceed the limit set by article KD-251.1 (division 3 of ASME BPVC section 8) given as:

$$P_D = \frac{1}{1.732}(Sy)\ \ln(Y) \tag{1}$$

$$137 = \frac{1}{1.732}(965)\ \ln(Y)$$

$$Y = 1.28$$

Where Y is the ratio of outer diameter (D_o) to the inner diameter (D_i) of the shell, $Y = D_o/D_i$, and ratio of 2 is assumed for safety. Assumed D_o =3048 mm, D_i=1524 mm, Length = 3 x D_o and flange thickness = 254 mm. The ellipsoidal top head is considered to be fully radiographed like shell, hence E = 1. Then using given equation:

$$t = \frac{PDk}{2SE - 0.2P} \tag{2}$$

Where K is the stress intensity factor by the equation below:

$$K = \frac{1}{6}\left[2 + \left(\frac{a}{b}\right)^2\right] \tag{3}$$

$$t = \frac{137 * 3048 * 1}{2 * 965 * 1 - 0.2 * 137}$$

Thickness = 220 mm

The required thickness of the bottom hemispherical head is normally one-half the thickness of an elliptical or torispherical head for the same design conditions, material, and diameter which will be 110 mm in this case.

For the openings and closures, which are an important requirement for safety, is given as:

$$t_{rn} = \frac{Pr}{SE - 0.6P} \tag{4}$$

$$t_{rn} = \frac{137 * 1524}{965 * 1 - 0.6 * 137}$$

Where r is the inside radius of nozzle, assuming a 1524 mm nozzle, the required thickness comes out to be 254 mm.

The CAD model of the pressure vessel is shown below:

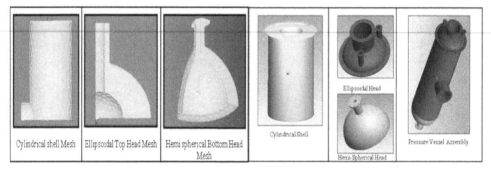

| Cylindrical shell Mesh | Ellipsoidal Top Head Mesh | Hemispherical Bottom Head Mesh |

Subfigure 1: Design model for case one

Subfigure 2: Meshing of pressure vessel components for case one

Fig. 3. Subfigure with two images

6.1.1 Element type

Each model part was meshed with SOLID 45 element with a higher concentration of elements around the nozzles and stress concentration areas. Eight nodes having three degrees of freedom at each node define the element translations in the nodal x, y, and z directions. The element has plasticity, creep, swelling, stress stiffening, large deflection, and large strain capabilities. Subfigure 2 of Fig.3 above shows the meshing of the model parts.

6.1.2 Boundary conditions

Axis symmetric displacement boundary conditions was applied to the two axes to ensure that the body, bottom end of the shell and the head are fully constrained with internal pressure loadings of 137 MPa and temperature of 400 oC applied. The finite analysis results are shown in tables 1 and 2 below.

Sections of Pressure Vessel	1st Principal Stress	Von Mises	Stress Intensity	Displacement
Cylindrical Shell	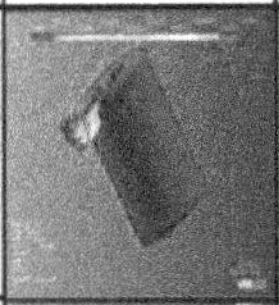	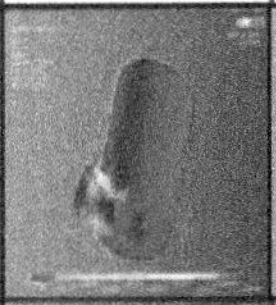		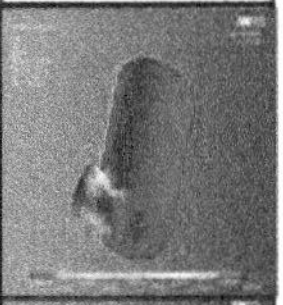
Ellipsoidal Head	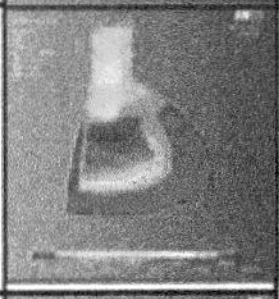	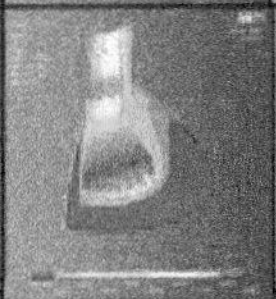		
Semi - Hemispherical Head				

Table 1. Result of finite analysis for design case one

Sections	1st Principal Stress (MPa)	Von Mises (MPa)	Stress Intensity (MPa)	Displacement (mm)
Cylindrical Shell	2562.59	2541.21	2552.94	0.38605
Ellipsoidal Head	704.35	623.94	720.15	0.09479
Semi Hemispherical Head	726.73	744.73	770.86	0.11313

Table 2. Summary of finite analysis result for design

6.2 Case two design model

$$P_D = \frac{1}{1.732}(Sy)\ \ln(Y)$$

After the finite element analysis of design case one, large stress was found around the nozzle junction of the cylindrical shell and also found stress developed at the bottom and top head. Ways of mitigating the stress is addressed in the model design case two. The design parameters are kept the same as in Design case one except for the temperature rating tensile strength of the material that is 806 MPa. The diametric ratio Y assumed is 1.5 derived from equation (1):

Outer and inner diameter assumed were 3048 mm and 2030 mm respectively. Reconsidering the nozzle thickness in design case two, the value of S is taken as 806 MPa using equation (4):

$$t_m = \frac{Pr}{SE - 0.6P}$$

Thickness = 152 mm

To mitigate the stress discovered during the finite element analysis of design case one, reinforcement pad is introduced to lower the stresses around the junction of shell and nozzle of the pressure vessel model. The total cross-sectional area of reinforcement A required in any given plane for a vessel under internal pressure should be not less than:

$$A = dt_r F \tag{5}$$

Where

d = diameter in the given plane of the finished opening, (mm).

t_r = minimum thickness which meets the requirements of KD-230 in the absence of the opening (mm).

And F is calculated from the graph in non-mandatory appendix H of section 8 division 3 of ASME BPVC to be 1.

Hence,

$$A = 1524 * 152 * 1$$

$$A = 9144\, mm^2$$

Using the formula for the area of circular disc given as,

$$A_{disk} = \pi(r_o^2 - r_i^2) \tag{6}$$

The outer diameter of the reinforcement pad = 1625 mm, while thickness = 101 mm. The top head is 2:1 ellipsoidal with thickness assumed calculated as 279 mm using equation (2):

$$t = \frac{PDk}{2SE - 0.2P}$$

Thick head with pressure on the concave side is used for the bottom hemispherical with given equation:

$$t = \frac{5 * 137 * 1524}{6 * 806} \tag{7}$$

$$t = \frac{5PL}{6S}$$

$$t = 216.9\, mm$$

6.2.1 Element type

SOLID 92 elements was used to mesh each model part in design case two, due to its quadratic displacement behaviour, its suitability to model irregular meshes and accuracy. Ten nodes having three degrees of freedom at each node define the elements: translations in the nodal x, y, and z directions. In addition, symmetric boundary condition was used; hence, half of the cylindrical shell and quarter of heads model were used for the analysis. The result of the finite simulation is shown in table 3 and summarised in table 4 below:

Sections of Pressure Vessel	1st Principal Stress	Von Mises	Stress Intensity	Strain
Cylindrical Shell				
Ellipsoidal Head				
Semi Hemispherical Head				

Table 3. Result of finite analysis for design case two

Summary of FEA Results	1st Principal Stress (MPa)	Von Mises (MPa)	Stress Intensity (MPa)	Displacement (mm)
Cylindrical Shell	792.31	695.74	780.40	0.15532
Ellipsoidal Head	2106.58	1777.62	1946.97	0.38747
Semi Hemispherical Head	2171.67	1653.04	1902.59	0.37863

Table 4. Summary of finite element results

6.3 Case three design model

In design case 3, the shortcomings of both design cases 1 and 2 were considered from design point of view to lower the stresses to an acceptable range. Reinforcement pad is used along with the use of diametric ratio of 2.0, i.e. Y = 2 as calculated in design case 1. Outer diameter of 3048 mm and the inner diameter of 1524 mm with length of cylindrical shell being three times the outer diameter for structure stability. Nozzle thickness kept same as design case 2.

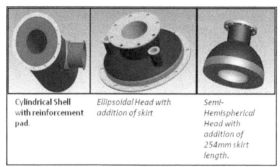

| Cylindrical Shell with reinforcement pad. | Ellipsoidal Head with addition of skirt | Semi-Hemispherical Head with addition of 254mm skirt length. |

Fig. 4. Design model for case three

6.3.1 Element type

Each component of the pressure vessel is simplified using symmetry. Solid 3D Element was used to mesh the models with fine mesh around the nozzle and shell junction. Pressure and temperature of 137 MPa and 400 °C applied on the internal surface of the components. The results obtained from finite analysis are given below:

Sections of Pressure Vessel	1st Principal Stress	Von Mises	Stress Intensity	Strain
Cylindrical Shell				
Ellipsoidal Head				
Semi Hemispherical Head				

Table 5. Result of finite analysis for design case three

Sections of Pressure Vessel	1st Principal Stress (MPa)	Von Mises (MPa)	Stress Intensity (MPa)	Strain (mm)
Cylindrical Shell	509.99	499.92	568.98	0.1132
Ellipsoidal Head	475.91	400.37	449.9	0.0895
Semi Hemispherical Head	659.11	567.5	567.5	0.1129

Table 6. Summary of finite element analysis result

7. Discussion

Finite element analysis result in design case one was compared with the allowable stress intensities of ASME BVPC Division 3 and it was found that the stresses in the cylindrical shell exceeded the value of the allowable stress with almost 70 %. Also, the result shows higher stress development along the nozzle length away from the shell and at the bottom semi spherical head. However, the stresses developed in ellipsoidal enclosures head has a very less margin than the allowable stress of about 4%, which can be improved by using better thickness to diameter ratio.

In the design case two model, the shell thickness was reduced and thickness of the nozzle increased as well as the addition of reinforcement pad of high alloy steel SA-705 grade XM-13 around the nozzle to shell junction. The finite element analysis shows that less stresses were developed in the shell wall than the allowable and there was a reduction of stress from 2552 MPa in design case one to 780 MPa in design case two; about 70% drop in stresses. However, despite the decline in the stresses to 780 MPa, it is still higher than the allowable stress by the ASME Code. Finally in the design case three, the cylindrical shell was modified and the reinforcement pad introduced in design case two retained to reduce stresses around the nozzle to shell junction and a skirt length of 254 mm was added around the enclosure heads. This is necessary due to the result of the analysis obtained in the design cases one and two which showed that highest stresses were concentrated on the lower end of the heads where there was formation of joins to the shell. The finite element analysis shows that with the addition of the skirt, there was lesser development of stresses in the lower ends of the heads. The table below shows the comparison of the stresses obtained from finite element analysis and the stresses allowed by the ASME Code.

Sections	Stress Intensity MPa		Allowable Stress MPa
	Min	Max	
Cylindrical Shell	1.42	568.9	747
Ellipsoidal Head	0.62	449.9	747
Semi Hemispherical Head	0.951	567.5	747

Table 7. Stress Intensity comparison of design case three

8. Conclusion

Three main factors are seen to contribute extensively to the development of stresses in pressure vessels. These are thickness, nozzle positions and the joints of the enclosure heads. From the model design cases used in this research, it could be seen that as the thickness of pressure vessel increases, the stresses decreases, however this is not a viable solution due to cost. Nozzles though are safety relief devices and important component of pressure vessels comes with its own disadvantages of increasing weak areas and stress concentration. However, this was mitigated by use of high alloy reinforcement pads as applied in the design case two and three of this work. The high strength reinforcement pad used has a chemical composition of titanium 0.4 to 1.20%; hollow disc shaped with rectangular section can also reduce the stresses concentration around the nozzle.

Finally, the joints of enclosure heads either welded or bolted were identified as areas with the highest concentration of stresses i.e. with peak stress. Addition of 254 mm skirt length at the end of enclosure heads provide more room for the stresses to develop slowly in the wall of the head regions, thus making the pressure vessels more resistant to the loadings.

9. References

AirConditioning,Heating,Refridgeration - The News (2002) . Boiler Accident Report : "To Err Is Human" .URL:http://www.achrnews.com

American Petroleum Institute (2001). Inspection of Pressure Vessels (Towers, Drums, Reactors, Heat Exchangers, and Condensers), Recommended Practice 572, Second Edition.

Bulk Transporter (2009) . "Boiler, Pressure Vessel Accidents Increase" . URL: http://bulktransporter.com

Chattopadhyay. S (2004): Overview of Pressure Vessels. Pressure Vessels Design and Practice. 3rd edition: CRC Press, pp. 1-10. United States of America.

D. R. Moss (2004). Pressure Vessel Design Manual – Illustrated Procedures for Solving Major Pressure Vessel Design Problems. 3rd Edition, Gulf Professional Publishing, Oxford, United Kingdom

Engineers Edge (2010) – Design for Manufacturability Reference and Training Book. Stress-Strength of Materials.
    URL: http://www.engineersedge.com.

Hearn.E.J. (1998). Mechanics of Materials, 2nd edition, Cambridge University Press, United Kingdom.

Megyesy. E.F (2001). Design of Tall Towers. Pressure Vessel Handbook. Pressure Vessel Publishing, Inc. United States of America.

Rao. K. R.(2002). Companion guide to the ASME boiler & pressure vessel code, ASME Press. United States of America.

Sharma. S. C. 2010. Strength of Materials. Department of Mechanical & Industrial Engineering. Indian Institute of Technology Roorkee. URL:
    http://nptel.iitm.ac.in/courses/Webcourse.

Sirosh. N & Niedzwiecki. A. (2008). Hydrogen Technology: Development of Storage Tanks - High Pressure Vessels, Springer- Verlag, Berlin Heidelberg.

Spence. J & Tooth.A.S. (1994). Pressure Vessel Design: Concepts and Principles. Taylor & Francis; 1st edition, London.

Syed U. A. (2009). Design and Analysis of Pressure Vessels under Elevated Temperature and Pressure. MSc Thesis, University of Teesside, United Kingdom.

The National Board of Boiler and Pressure Vessel Inspectors (2002). National Board Bulletin. URL: http://www.nationalboard.org

5

Finite Element Analysis of Multi-Stable Structures

Fuhong Dai* and Hao Li
Center for Composite Materials and Structures,
Harbin Institute of Technology,
China

1. Introduction

1.1 Conception of multi-stable structures

A structure that has at least two stable geometric configurations is called multi-stable. The critical load level has to be reached before snap-through occurs, which means the stable states can carry load. The multi-stable structure is a system owning more than two stable configurations. The first important aspect is that large shape change can be accomplished with small energy input and without complicated actuators. Instead of the power being supplied to elastically deform over its entire range, power is only needed to snap the structure from one stable configuration to another. A reduction in weight of the overall structure is possible, since the whole structure can serve as both the base structure and the control surface.

It is receiving the attention of scientists in developing the multi-stable structure which enables a number of operational shapes (Hufenbach et al., 2002; Portela et al., 2008). It is potentially suitable for a wide variety of systems, such as morphing aircraft (Yokozeki et al., 2006; Diaconu et al., 2008; Schultz, 2008), deployable structures (Lei & Yao, 2010) and mechanical switches. The studies of multi-stable structures reported in open literatures are dominated by the bi-stable unsymmetric composite laminates (Jun & Hong, 1992; Dano &

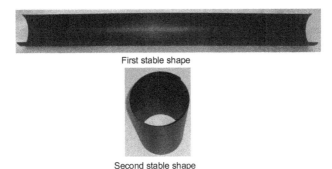

First stable shape

Second stable shape

Fig. 1. The bi-stable unsymmetric composite laminate

---

* Corresponding Author

Hyer, 1998; Maenghyo et al.,1998; Hufenbach et al., 2002, 2006; Ren, & Parvizi-Majidi, 2003, 2006; Gigliotti et al., 2004; Dai et al. 2007; Etches etal., 2009). A typical bi-stable composite laminate is shown as in Fig.1, which means that applying an external force the room temperature cylindrical shape can be snapped into another cylinder with curvature of equal magnitude and opposite sign.

1.2 Design and analysis method

How to design and analyse the multi-stable behaviour is still a challenging problem. It is actually a bifurcation problem. First issue about the multi-stable structures is to predict their multiple shape configurations. Second important issue is to simulate the snap-through behaviors of multi-stable structures.

1.2.1 Analytical method

Hyer observed the large deformation of thin [0/90] cross-ply unsymmetric composite laminates in curing experiments (Hyer, 1981). They found that the shapes of cured unsymmetric composite laminates were a part of cylinder, which were stable at room temperature. To take into account the large multi-stable deformations, which are often many times over the laminate thickness, the linear strain–displacement relations must be extended by non-linear terms. Since then, many researchers had employed the Rayleigh–Ritz approach proposed by Hyer to predict the shape of unsymmetric laminates (Hyer, 1982; Maenghyo & Hee, 2003). Jun (Jun & Hong, 1992) modified Hyer's theory by including more terms in the polynomials to take plane shear strain into account. Dano (Dano & Hyer, 1998) used a higher-order polynomial to calculate the displacement field. Dai and Daynes (Dai etal., 2009; Daynes & Weaver 2010) predicted the cured shape of the bi-stable hybrid composite laminate. There is a very good agreement between the experimental and analytical shapes predicted by Rayleigh–Ritz method. The cured shape and bifurcation point can be obtained. An example of bistable analytical solutions is shown as in Fig.2 (Dai etal., 2009).It can be seen that there only exist one group of solutions when the side length is less than some critical value (35 mm), as the branch curve AB and A'B' plotted in Fig.2. Here the curvature in x-direction is opposite to that in y-direction. There exist three groups of solutions when the side length is more than some critical value, as shown with the branch

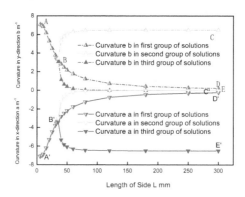

Fig. 2. An example of analytical solution curves for bi-stable composite laminate

curves of BD and B'D', BC and B'C', BE and B'E'. The first group of solutions as shown with the branch curve of BD and B'D' mean a saddle shape, which is not in reality. The second and third group of solutions mean that snap-through phenomenon occurs. The curvature of curve BC is 6.49 and the corresponding value of curve B'C' is approximately 0. It indicates the cured shape of unsymmetric hybrid composite laminates is similar to a half cylinder.

1.2.2 Finite element method

In addition to analytical method, Finite Element Analysis (FEA) is found to be a robust way to study the bi-stable laminates. Snap through analysis of multi-stable structures requires the solution of a nonlinear system of equations. The iterative techniques are used to solve these equations, where small incremental changes in displacement are found by imposing small incremental changes in load on the structure. The arc length methods were commonly used to overcome the problem of tracing the equilibrium path in the neighborhood of limit points (Riks, 1979; Wempner, 1971).Schlecht employed the FEA to predict the cured shape of unsymmetric laminates with the commercial finite element code Marc (Schlecht et al., 1995,1999). Giddings combined a finite element model of a Macro-Fiber Composite (MFC) with a bi-stable asymmetric laminate model. Both predicted shape and snap-through voltage of a piezo-actuated [0/90] laminate agreed well with experimental results (Giddings et al.m 2011)).In recent years, another commercial finite element code ABAQUS is widely employed, which is also accurate enough in shape prediction of bi-stable laminates (Dai,2011; Mattioni,2008,2009; Portela et al., 2008). ABAQUS offers two procedures, namely RIKS and STABILIZE, both are capable of solving a nonlinear system of equations and hence suitable procedures for studying unstable post-buckling problems. The formulation behind each procedure is rather different and hence, requires different sensitivity of their respective input parameters. Tawfik systematically discussed the two methods (Tawfik et al., 2007), namely "RIKS" and "Stabilize", both were capable of studying the snap-through of bi-stable laminates. Mattioni pointed out that the "Stabilize" algorithm is more suitable (Mattioni, 2008, 2009). Portela studied the snap-through of laminate with actuator (MFC) by FEA (Portela et al., 2008).

2. Finite Element Analysis of bi-stable laminates

There are two Finite Element Methods to solve the bi-stable bifurcation problem: one-step method and two-step method. In one-step method, the geometrical model without internal thermal stress is directly built, which is actually a stable structure. Then the loads are applied on the structure and the critical load can be solved. In two-step method, the cool-down process is firstly simulated and the room temperature bi-stable geometrical models are built. Then, the snap-through simulation of the stable configurations can be conducted. The following section gives the detail examples for the both models.

2.1 Experiments of bi-stable laminates

The experimental set up is given as follows before performing finite element analysis. The testing specimen is [0/90] cross-ply epoxy matrix carbon fiber-reinforced composite laminate with the size of 140mmx140mm. The materials in the experiments are T300/Epoxy prepreg with longitudinal elastic modulus E_1 of 137.47 GPa, transverse elastic modulus E_2 of

10.07 GPa, Possion's ratio v_{12} of 0.23, the longitudinal linear expansion α_1 of $0.37\times10^{-6}/°C$ and transverse linear expansion α_2 of $24.91\times10^{-6}/°C$.

As shown in Fig.3, the composite laminate is put on a pair of slippery orbits to ensure that the laminate is supported only at four corner points. A concentrated force is applied on the geometrical center of the laminate, by a special loading apparatus which can increase the load gradually. The concentrated force is gradually increased to the critical load of snap-through and the height of the center point at every step is measured.

Fig. 3. Schematics of concentrated force load experiments

The load-displacement curve is shown in Fig.4. The displacement denotes the deflection at the middle of laminate's span. It can be observed that the curve is almost straight up at the beginning, and the load reaches its maximum at the end of curve, which is the critical load of snap-through. The critical load is 1.32N and the corresponding displacement at the center of the laminate is 21mm, which denotes the critical deformation of snap-through. Because of the limitation of loading apparatus, the unloading part of the curve is not obtained.

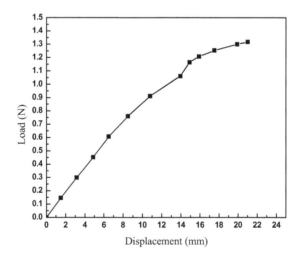

Fig. 4. Load-displacement curve of bi-stable laminates

2.2 Two-step FEA method

In this section, the commercial finite element code ABAQUS is employed to simulate the snap-through of bi-stable laminate, and the predicted results such as critical load and deformation of the composite laminate are presented.

The first step is to simulate the cool-down process of composite laminate with internal thermal stress. Although the practical composite laminate is a square plate with the side length of 140mm, a geometrical imperfection in the finite element model should be introduced. Otherwise the simulation result will be a saddle-shape rather than the expected cylindrical shape. The cool-down process in practice is time-consuming (a few hours depending on the curing cycle that the material requires). The first analysis step is "Static". To perform the geometrical non-linear analysis, the option "NLGEOM" is employed. The element type is S4R and the center node of the FEA model is fully constrained. During the analysis, the model is cooled down from 160℃ to 20 ℃, and the deformation is merely generated by the temperature change, then first stable shape can be obtained. To get another stable shape, the "Static, Stabilized" step with an automatic stabilization which is based on the addition of viscous forces to the global equilibrium equation is used and the solution scheme is able to converge to another stable shape.

The simulation results and the measured results are listed in Table 1. It can be seen that the numerical results agree well with the measured results.

	Experimental	Numerical	Error (%)
Deflection (mm)	25.90	27.268	--
Span (mm)	126.3	126.06	--
K (m^{-1})	11.19	11.49	2.64

Table 1. Experimental results and FEA results of cured bi-stable laminates

After the cool-down simulation, the temperature is kept at 20℃ and the following boundary condition and displacement load are applied on the model.

1. The center node of the laminate is fully constrained.
2. Four same displacement loads are applied on the four corner nodes, parallel to the vertical axis, pointing to the opposite direction of the corner nodes' current position.

The "Static, Stabilize" step is employed here, under the boundary conditions above, the reaction force of the center node is the corresponding center load in experiment. The simulation process is shown in Fig-5.

The curve of the reaction force of center node versus its relative displacement is presented in Fig-6.It is observed that the FEA results are greater than the experimental and the curve is more complex. The error of critical load is about 28% since finite element model is an ideal model while the specimen contains some defects which are generated in the manufacturing process. In FEA, the thickness of the model is 0.25mm, which is equal to the sum of two plies. However, the thickness of the specimen is not equal to 0.25mm exactly. The actual thickness is about 0.23±0.02mm. Therefore the modified thickness 0.23mm is used to calculate again. The results are shown in Fig.6. The results of modified model are much improved and agree well with experiment. The calculated critical load is 1.33N, and the error of critical load decreases to 0.7%.

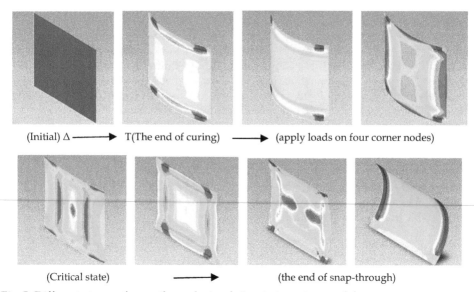

(Initial) Δ ——————▶ T(The end of curing) ——————▶ (apply loads on four corner nodes)

(Critical state) ——————▶ (the end of snap-through)

Fig. 5. Different stages of snap-through simulation in two-step model

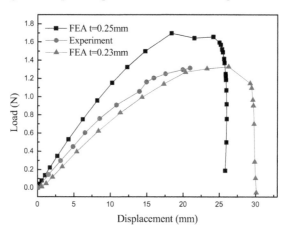

Fig. 6. Load-displacement curves of numerical simulation and experiment

2.3 One-step FEA method

Here, a part of cylindrical model with no internal thermal stress is directly built. The initial state of the model has the same geometrical configuration as the specimen in experiment, such as dimensions and curvature. The thickness is chosen to be 0.23mm. The snap-through process is shown in Fig.7.

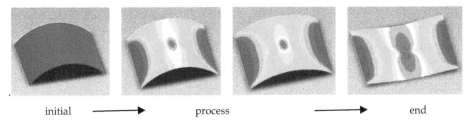

initial ——————▶ process ——————▶ end

Fig. 7. Different stages of snap-through simulation in one-step model

The results of one-step model are compared with two-step model and experiments. In Fig.8, it can be observed that the curve of one-step model is almost linear and has no critical point. Interestingly, the load-displacement curves of one-step model and two-step model almost agree with each other when the deformation is small, however, as the displacement increases, the difference between the two kinds of models increases rapidly. Thus, to study an unsymmetric composite laminate, when the deformation is small, it is more convenient to directly build a model with the same curve, and the results will be accurate enough.

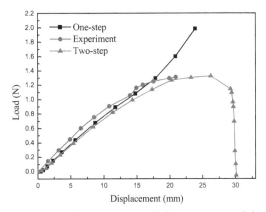

Fig. 8. Load-displacement curves in one-step model, two-step model and experiment

3. Snap-through FEA of multi-stable lattice structures

The studies of multi-stable structures are dominated by the bi-stable unsymmetric composite laminates. In some applications there may be a need to fabricate multi-stable structures with more than 2 stable configurations. One of important issue is how to fabricate the multi-stable structures. The additional constraint and the change of stresses distribution may cause the loss of multi-stable capability when the bi-stable laminates are bonded to other structural components. Here, the tri-stable lattice structures based on bi-stable laminates are presented and some simulation studies are given.

3.1 Design and fabrication of tri-stable lattice structures

The lattice structure is made up by four same rectangular composite laminates with the size of 140mm×35mm×0.25mm. The stacking sequence is [0/90]. Due to the fact that the size of laminates is too small, it is not convenient to manufacture the laminates. In experiment, we first manufactured a square laminate with the size of 140mmx140mmx0.25mm. Then the square laminate is divided into four same rectangular laminates. The rectangular laminates have two stable configurations, namely configuration A and B in Fig.9. The lattice element is composed of four narrow rectangular bi-stable laminates. The measured average curvature of rectangular laminates is 8.85m⁻¹.

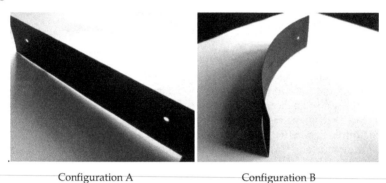

Configuration A   Configuration B

Fig. 9. Two stable configurations of rectangular cross-ply composite laminate

The four rectangular laminates are connected by the bolts. The narrow rectangular laminate has the same curvature in both the longitudinal and transvers directions. Thus it can offer the smooth and close contact between two laminates when the narrow rectangular laminate is bent along the longitudinal direction. The lattice structure, as presented in Fig.10, has three stable configurations without external constraint. The first stable shape is like a plane rectangular lattice. The second stable shape is like a concave lattice. The third stable shape is like a convex lattice.

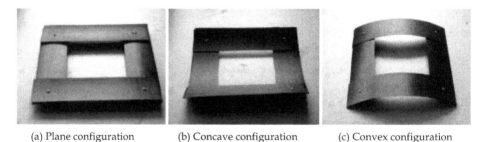

(a) Plane configuration   (b) Concave configuration   (c) Convex configuration

Fig. 10. Three stable configurations of lattice structures

3.2 Experiments of tri-stable lattice structures

The concave configuration is unable to directly transform to convex configuration. It has to transform to the plane configuration first, then to the convex configuration. So we can just investigate the snap-through between the plane configuration and the two curving configurations. We name the plane as configuration A, and name the convex configuration as configuration B, as seen in Fig.11 and 12. For loading scheme of configuration A, A concentrated force is applied on the center of rectangular laminate, and the two supports are set symmetrically. For the lattice structure, two same concentrated loads are applied simultaneously on the two centers of rectangular laminates and the four supports are set symmetrically as well. The critical loads to snap through the rectangular laminate and the lattice structure are measured. The relationship between the critical loads and the position of supports is also investigated by changing the distance between the load and supports. For loading scheme of configuration B, the specimens are put on a slippery glass, and a concentrated load is also applied on the centers of rectangular laminates.

Fig. 11. The loading scheme for configuration A

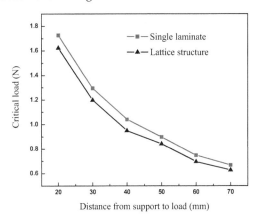

Fig. 12. The loading scheme for configuration B

Fig. 13. Measured average critical loads of configuration A of both separate laminates and lattice structure

For configuration A, when load slowly increases to the critical load, the rectangular laminates quickly snap from configuration A to B. Because of some unpredictable defects introduced in manufacture process, the mechanical properties of four rectangular laminates vary from each other, which give rise to difference of their critical loads. The average measured critical loads for configuration A are presented in Fig.13. In order to compare with the separate rectangular laminates, half of the critical loads of lattice structure are given. The experimental results reveal that the critical load increases as the distance from support to load decreases and the relationship is not linear.

For configuration B, the experimental phenomenon is much different from configuration A. The deformations of separate laminates and lattice structures before snap-through are much greater than those of configuration A while the critical loads are much smaller. Under the loading scheme of configuration B, the measured critical loads of four separate laminates are presented in Table 1, and the measured critical loads of lattice structure are presented in Table 2. It shows that the average critical load of lattice structure, divided by 2, 0.177N, is 6.8% smaller than the average critical loads of separate laminates, 0.190N.

Lattice structure	Test 1	Test 2	Test 3	Test 4	Average
Critical load(N)	0.361	0.348	0.356	0.351	0.354

Table 2. Measured critical load of lattice structure for configuration B

3.3 Snap through FEA of tri-stable structures

The commercial finite element code ABAQUS is used to simulate the snap-through of rectangular laminates and lattice structure. The parameters used in the simulation are same as the above. A two-step method is proposed. First, it is required to simulate the cool-down process to build the initial finite element models of laminates and lattice structure. It often takes hours for the laminates to cool down from the curing temperature to room temperature. So the cool-down process is thought to be a quasi-static process. In the simulation, the laminate is modeled using four-node-square shell elements (S4R). The four bi-stable laminates are numerically generated during cool-down process by applying an initial curing temperature and a final room temperature. For lattice structures, the four components are bonded together before the cool-down simulation by using four "tie" constraints in ABAQUS, which constrain the relative displacements of the bonding points. Calculated curvature and measured curvatures of separate laminates are listed in Table 3.

Specimen	1	2	3	4	average
Measured curvature m⁻¹	8.44	8.70	9.21	9.06	8.85
Simulated curvature m⁻¹			9.18		
Error (%)	8.77	5.52	0.33	1.32	3.73

Table 3. Simulated and measured curvatures

In second step, after the cool-down simulation, the snap-through simulation of the stable configurations can be conducted. The snap-through problem of unsymmetric laminates is a buckling problem in essence, which can be unstable. However, if the instability is localized, there will be a local transfer of strain energy from one region of the model to neighboring region, and global solution methods may not work. This problem has to be solved either dynamically or with the aid of artificial damping. Abaqus/Standard (Simulia 2009) provides an automatic mechanism for stabilizing unstable quasi-static problems through the automatic addition of volume-proportional damping to the model. To ensure an accurate solution is obtained with automatic stabilization, the ratio between the viscous damping energy and the total strain energy should not exceed the defined dissipated energy fraction or any reasonable amount. Fig.14 presents finite element models and the loading illustration for configuration A, which is used to simulate the snap-through and to compare with the experiments. According to loads and boundary conditions in experiments, a concentrated force is applied on the center point of finite element model, and vertical movements at supports are constrained. For lattice structures, two same concentrated forces are applied simultaneous on center points of two parts, vertical movements at supports are also constrained.

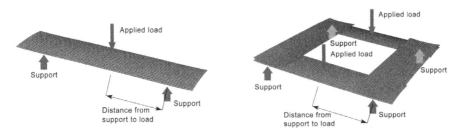

Fig. 14. Illustration of finite element model and loading scheme for configuration A

For configuration B, the snap-through simulation is a little more complicated, and it is required to build an analytical plane to simulate the glass plane in experiments. The surface to surface contact between the model of laminate and the analytical plane should be established to simulate the interaction between the specimens and the glass. Since the critical load of configuration B is very small, the gravity is considered here. The loading scheme is illustrated in Fig.15. In the models of both single laminate and lattice structure, concentrated forces are applied on the center points, and vertical movements at supports are constrained.

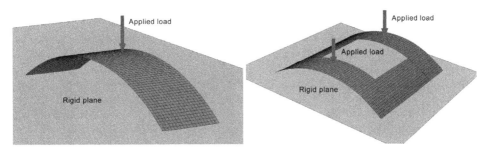

Fig. 15. Illustration of finite model and loading scheme for configuration B

In experiments, the specimens are snapped too fast to observe the process clearly, but through the FEA we can observe every step of snap-through under the presented loading scheme. As shown in Fig.16, for configuration A, when loads increase to certain value, local instability is observed firstly in the center of model and snap-through starts from there. Regions between snapped and unsnapped regions turn to be unstable, then unstable regions gradually move towards short edges and snapped region spread. During the process of snap-through, unstable regions move smoothly toward short edge. When they get to short edges, the plane configuration is snapped to be a convex configuration.

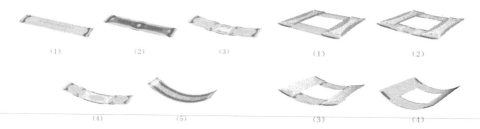

Fig. 16. Simulated snap-through process for configuration A

Because the finite element model is ideal, the loads applied on the two components of lattice structures are the same. The applied load-displacement curve of configuration A according to the different positions of supports is shown in Fig.17. The peak points of the curvatures are the critical loads of snap-through for loading scheme of configuration A. When the load increases to critical load, central regions lose stability

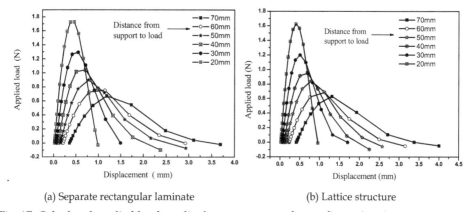

(a) Separate rectangular laminate   (b) Lattice structure

Fig. 17. Calculated applied load vs. displacement curves for configuration A

Under the loading scheme illustrated in Fig.15, the snap-through process of configuration B given by FEA is presented in Fig.18. The curvatures decrease with increasing loads. When the curvatures decrease to the critical value, and then center region starts to lose stability and snap-through also starts from the center region. Snap-through spreads gradually form the center region towards toward two short edges. Finally, the lattice structure is snapped to be a plane configuration when the unstable region spread to short edges.

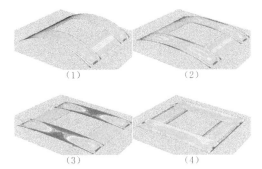

Fig. 18. Simulated snap-through process for configuration B

The calculated load-displacement curves of configuration B are given in Fig.19. The peak points of the two curves in Fig.13 are critical points of snap-through. As same as experiments, the calculated critical load of lattice structure is a little smaller than separate laminates, 0.1836N and 0.1942N respectively.

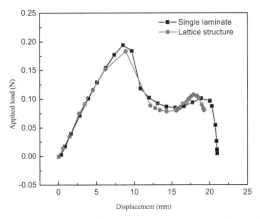

Fig. 19. Calculated applied load vs. displacement curves for configuration B

Compared Fig.17 with Fig.19, it can be found that the deformation of configuration A before snap-through is smaller than that of configuration B, however the critical load of configuration A is much greater than that of configuration B. The lattice structure is made up by the four rectangular composite laminates, so its mechanical behaviours are mainly dominated by the mechanical behaviours of rectangular laminates. Since the critical loads of four components of lattice structure are different, the two components may not snap at the same time even under symmetric loads. The part with relative smaller critical load will snap first. Apparently, the slower the load increases, the greater the time lag is, this will weaken the continuity of snap-through process of lattice structure. It is needed to find a better way to snap the lattice structure more continuous as an integral structure instead of tied parts.

The two configurations of rectangular bi-stable composite laminates curl along long edge and short edge respectively, and respond differently under centre load. In Fig.16 and Fig.18, it is found that both configurations snap locally in center first. In Fig.17, it shows that before local snap the deformation is very small. In Fig.19, it shows the critical load is much smaller, and the deformation is much greater, and after the critical point the applied load does not decrease to zero as configuration A. The laminate has a displacement jump on a certain load then the load decreases to zero as snapped region spreads to the entire laminate. FEA results denote that the bending stiffness of configuration A is greater. The error between calculated curvature and measured curvature is 3.73%. It indicates that the initial finite element models are reasonable. The comparisons of critical loads of configuration A between FEA and experiments are given in Fig.14. The calculated critical loads under the same loading scheme are about 11-18% smaller than measured values. For configuration B, The calculated results are greater than the measured values. The measured critical load of separate rectangular laminate is 0.190N, while the calculated is 0.194N. The measured critical load of lattice structures is 0.177N, while the calculated is 0.183N. The error is about 3%~4%.

The main reason giving rise to errors of calculated critical loads is that the finite element models are ideal while the specimens inevitably contain some defects, in addition, complex boundary constraints may be not fully characterized in numerical models. How to consider the defects and the constrain effects in the snap-through simulation is needed to be studied further.

4. Conclusions

The finite element analysis based on general purpose software of ABAQUS was successfully used to simulate the snap-through behaviors of multi-stable structures. The two-step model and the one-step model were introduced to predict the critical load. The results from two-step model show a good agreement with experiment, the error of critical load decreases from 28% (ideal model) to 0.7% (modified model). The results of one-step model show that the model without internal thermal stress is not bi-stable. However, the comparison between the two kinds of models indicates that when the deformation is relative small (about half of the critical deformation), the results of the two kinds of models agreed well. Therefore, to study an unsymmetric bi-stable composite laminate, when the deformation is small, it is more convenient to directly build a model with the same curvature, and the results will be accurate enough. Otherwise, the model with internal thermal stress must be considered.

It is undoubtedly very important and requisite in engineering how to fabricate multi-stable (more than 2) structures and to get different stable configurations. A method for assembling the bi-stable laminates to fabricate a multi-stable structure has been presented. The snap-through of rectangular cross-ply composite laminates and a lattice structure was numerically and experimentally studied. The experiments to measure the snapping force levels of rectangular laminates and lattice structures were conducted. A two-step method based on commercial software ABAQUS to simulate the snap-through behaviors of lattice structures were presented. Snap through of lattice structures were successfully described with the use of this method. The same phenomenon between the FEA and experiments were observed. The predicted curvatures show a good agreement with experimental results while there are some errors between predicted and measured critical loads. The errors of critical loads are caused by unpredictable defects and complex boundary constrains. How to consider the defects and the constrain effects in the snap-through simulation is needed to be studied further. The proposed tri-stable structures and two-step method are instructive to construct the multi-stable structures and to understand their snap-through behaviors. Finally, It should be noted that deep understanding of their behaviours will need to await more detailed studies, for example, the reliability and the fatigue properties, et al.

5. Acknowledgments

Thanks to support by National Natural Science Foundation of China (Grant No.10872058) and the Major State Basic Research Development Program of China (973 Program) under grant No. 2010CB631100

6. References

Dai, FH., Zhang, BM., He, XD. & Du, SY. (2007). Numerical and Experimental Studies on Cured Shape of Thin Unsymmetric Carbon/Epoxy Laminates. Key Engineering Materials, 334–335: 137–140

Dai, FH., Li, H. & Du, SY. (2009). Cured shape prediction of the bi-stable hybrid composite laminate. Proceeding of SPIE, 2009, 7493:1–7

Dai, FH., Li H., (2011). Prediction of Critical Center Load for Bi-stable Laminates. Polymers & Polymer Composites ,19 (2-3): 171-175 2011

Dano,ML. & Hyer, MW. (1998).Thermally-induced deformation behavior of unsymmetric laminates.International Journal of Solids and Structures, 35,2101-2113

Daynes, S. & Weaver, P. (2010). Analysis of unsymmetric CFRP–metal hybrid laminates for use in adaptive structures. Composites Part A, 41:1712–1718

Diaconu, CG., Weaver, PM. & Mattioni, F. (2008). Concepts for morphing airfoil sections using bi-stable laminated composite structures. Thin-Walled Structures 46:689–701

Etches, J., Potter, K., Weaver, P. & Bond, L. (2009). Environmental effects on thermally induced multistability in unsymmetric composite laminates. Composites Part A, 40:1240¬1247

Giddings, PF., Kim, HA., Salo, AIT. & Bowen, CR. (2011). Modelling of piezoelectrically actuated bi-stable composites. Materials Letters, 65:1261–1263

Gigliotti,M., Wisnom,MR.& Potter,KD.(2004). Loss of bifurcation and multiple shapes of thin [0/90] unsymmetric composite plates subject to thermal stress. Composites Science and Technology. 64,109 – 128

Hufenbach, W., Gude, M. & Kroll, L. (2002). Design of multi-stable composites for application in adaptive structures. Composites Science and Technology, 62:2201–2207

Hufenbach, W., Gude, M. & Czulak,A. (2006). Actor-initiated snap-through of unsymmetric composites with multiple deformation states, Journal of Materials Processing Technology. 175, 225–230

Hyer, MW. (1981). Some Observations on the Cured Shape of Thin Unsymmetric Laminates. Journal of Composite Materials, 15 (3): 175–194

Hyer, MW. (1982).The room-temperature shapes of four layer unsymmetric cross-ply laminates. Journal of Composite Materials, 16 (5): 318–340.

Jun, WJ. & Hong, CS. (1992). Cured Shape of Unsymmetric Laminates with Arbitrary Lay-Up Angles. Journal of Reinforced Plastics and Composites, 11: 1352–1365

Lei, YM. & Yao, XF. (2010). Experimental Study of Bi-stable Behaviors of Deployable Composite Structure. Journal of Reinforced Plastics and Composites, 29(6):865-873

Maenghyo, C., Min-Ho, K. & Heung, SC. (1998). A Study on the Room-Temperature Curvature Shapes of Unsymmetric Laminates Including Slippage Effects. Journal of Composite matrials, 32:460-482

Maenghyo, C. & Hee, YR. (2003). Non-linear analysis of the curved shapes of unsymmetric laminates accounting for slippage effects. Composites science and technology, 63: 2265-2275

Mattioni,F., Weaver, PM., Potter, KD. & Friswell, MI. (2008).Analysis of thermally induced multistable composites. International Journal of Solids and Structures, 45(2):657–675

Mattioni, F., Weaver, PM. & Friswell, MI. (2009). Multi-stable composite plates with piecewise variation of lay-up in the planform. International Journal of Solids and Structures, 46:151-164

Portela, P., Camanho, P., Weaver, P. & Bond, L. (2008). Analysis of morphing, multi stable structures actuated by piezoelectric patches. Computers and Structures, 86:347-356

Ren, LB., Parvizi-Majidi, A. & Li ZN. (2003). Cured Shape of Cross-Ply Composite Thin Shells. Journal of Composite Materials, 37(20): 1801–1820

Ren, LB., Parvizi-Majidi. A.(2006). A Model for Shape Control of Cross-ply Laminated Shells using a Piezoelectric Actuator. Journal of Composite Materials, 40(14):1271–1285.

Riks, E. (1979). An Incremental Approach to the Solution of Snapping and Buckling Problems, International Journal of Solids and Structures, 15: 529 – 551

Schlecht, M., Schulte, K. & Hyer, MW. (1995). Advanced calculation of the room-temperature shapes of thin unsymmetric composite laminates. Composite Structure, 32: 627–633

Schlecht, M. & Schulte, K. (1999). Advanced Calculation of the Room-Temperature Shapes of Unsymmetric Laminates. Journal of Composite Materials, 33(16): 1472–1490

Schultz, MR. (2008). Concept for Airfoil-like Active Bi-stable Twisting Structures. Journal of Intelligent Material Systems and Structures, 19(2):157–169

Simulia (2009). ABAQUS Analysis User's Manual 7.1.1, Solving nonlinear problems

Tawfik, S., Tan, XY., Ozbay, S. & Armanios, E. (2007).Anticlastic Stability Modeling for Cross-ply Composites. Journal of Composite Materials, 41(11):1325–1338.

Wempner, G.A. (1971). Discrete Approximations Related to Nonlinear Theories of Solids,International Journal of Solids and Structures, 7: 1581 – 1599

Yokozeki, Tomohiro M., Takeda, Shin-ichi., Ogasawara,Toshio. & Ishikawa,Takashi. (2006). Mechanical properties of corrugated composites for candidate materials of flexible wing structures. Composites: Part A. 37, 1578–1586

6

Application of Finite Element Analysis in Sheet Material Joining

Xiaocong He
Kunming University of Science and Technology,
PR China

1. Introduction

Lightweight construction strategies have become increasingly important in recent years for economic reasons and in terms of environmental protection. Some sheet material joining techniques, such as self-pierce riveting, mechanical clinching and structural adhesive bonding, have been developed for joining advanced lightweight sheet materials that are dissimilar, coated, and hard to weld (He et al., 2008, He, 2010b, 2011b, 2012).

Traditionally, the mechanical behavior of a sheet material joint can be obtained by closed-form equations or experiments. For a fast and easy answer, a closed-form analysis is appropriate. However, the mechanical behavior of sheet material joints is not only influenced by the geometry of the joints but also by different boundary conditions. The increasing complex joint geometry and its three-dimensional nature combine to increase the difficulty of obtaining an overall system of governing equations for predicting the mechanical properties of sheet material joints. In addition, material non-linearity due to plastic behavior is difficult to incorporate because the analysis becomes very complex in the mathematical formulation. The experiments are often time consuming and costly. To overcome these problems, the finite element analysis (FEA) is frequently used in sheet material joints in recent years. The FEA has the great advantage that the mechanical properties in a sheet material joint of almost any geometrical shape under various load conditions can be determined.

For having a knowledge of the recent progress in FEA of the sheet material joints, latest literature relating to FEA of sheet material joints is reviewed in this chapter, in terms of process, strength, vibration characteristics and assembly dimensional prediction of sheet material joints. Some important numerical issues are discussed, including material modeling, meshing procedure and failure criteria. It is concluded that the FEA of sheet material joints will help future applications of sheet material joining by allowing system parameters to be selected to give as large a process window as possible for successful joint manufacture. This will allow many tests to be simulated that would currently take too long to perform or be prohibitively expensive in practice, such as modifications to joint geometry or material properties. The main goal of the chapter is to review recent progress in FEA of sheet material joining and to provide a basis for further research.

2. FEA of self-pierce riveting

Self-pierce riveting (SPR) is used for high speed mechanical fastening of sheet material components. In this process, a semi-tubular rivet is pressed by a punch into two or more substrates of materials that are supported on a die. The die shape causes the rivet to flare inside the bottom sheet to form a mechanical interlock as shown in Fig. 1. Fig. 2 shows the SPR machine and the SPR tools in Innovative Manufacturing Research Centre of Kunming University of Science and Technology.

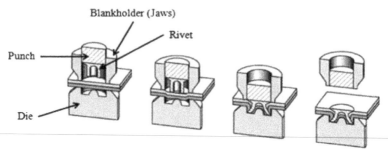

Fig. 1. SPR operation with a semi-tubular rivet.

Fig. 2. SPR machine and SPR tools.

2.1 SPR process

It is very difficult to get insight into the joint during forming process due to the complicacy of the SPR process. The effective way to analyze SPR joint during forming process is to perform finite element simulation. Several numerical techniques and different FEA software already allows the simulation of the SPR process.

The riveting process has been numerically simulated using LS-DYNA (Yan et al., 2011). A 2D axi-symmetric model was used with an implicit solution technique encompassing 'r-adaptivity' and geometrical failure based on the change in thickness of the substrates. An extensive experimental program using aluminum alloy 5052 substrates generated a database for the validation of the numerical simulations. Good agreements between the simulations and laboratory test results were obtained, for both the force/deformation curves and the deformed shape of the rivet and substrates. Fig. 3 shows the FE simulation of the SPR process and Fig. 4 shows the cross section comparison between simulation and test.

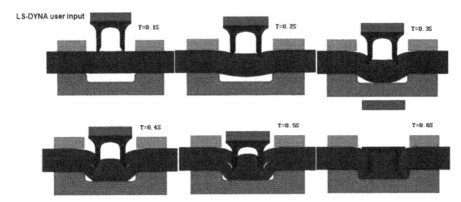

Fig. 3. FE simulation of the SPR process (Yan et al., 2011).

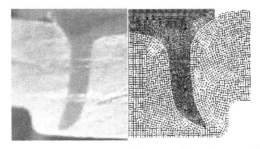

Fig. 4. Cross-section comparison between simulation and test (Yan et al., 2011).

A study (Abe et al., 2006) investigated the joinability of aluminum alloy and mild steel with numerical and test results. The joining process was simulated by LS-DYNA 2D axi-symmetric models. To shorten the calculation time for this explicit dynamic FEA code, the punch velocity was increased to about 25 times the real speed used for riveting. In this study, defects in the process were categorized as either penetration through the lower sheet, necking of the lower sheet or the separation of substrates (caused by different strains when substrates are made of different materials). The authors concluded that these defects were caused by the small total thickness, the small thickness of lower sheet and the large total thickness, respectively.

Mori et al. (Mori et al., 2006) developed an SPR process for joining ultra-high-strength steel and aluminum alloy sheets. To attain better joining quality, the die shape was optimized by means of the FEA without changing mechanical properties of the rivet. Authors reported that the joint strength is greatly influenced by not only the strength of the sheets and rivets but also the ratio of the thickness of the lower sheet to the total thickness. Abe et al. (Abe et al., 2009) investigated the effects of the flow stress of the high-strength steel sheets and the combination of the sheets on the joinability of the sheets by FEA and an experiment. They found that as the tensile strength of the high-strength steel sheet increases, the interlock for the upper high-strength steel sheet increases due to the increase in flaring during the driving through the upper sheet, whereas that for the lower high-strength steel sheet decreases.

A 3D model was created by Atzeni et al. (Atzeni et al., 2007) in ABAQUS Explicit 6.4 and both the SPR process and the shear tests were simulated to take into account the strain and residual stress of the SPR joints. Comparisons with experimental results had shown good agreement, both in terms of deformed geometry and force-displacement curves. In another study, the same authors (Atzeni et al., 2009) presented a FEA model for the analysis of the SPR processes. Correct model parameters were identified and numerical model validated in 2D simulations. In order to verify the capabilities of the software to predict joint resistance for given geometry and material properties, a 3D model was set up to generate a joint numerical model for simulating shearing tests. In Kato et al.'s paper (Kato et al., 2007), a SPR process of three aluminum alloy sheets was simulated using LS-DYNA to find joinable conditions. In addition, the cross-tension test was also simulated by FEA to evaluate the joint strength.

Using Forge2005® FE software, Bouchard et al. (Bouchard et al., 2008) modeled large deformation of elastic–plastic materials for 2D and 3D configurations. They found that it is possible to export the mechanical fields of a 2D simulation onto a 3D mesh using an interpolation technique, and then to perform a 3D shearing test on the riveted structure. They also found that the mechanical history of the rivet/sheet assembly undergone during the SPR process plays a significant role in the numerical prediction of the final strength of the assembly. In order to evaluate the software robustness, numerical simulation of the SPR process was performed on three 1-mm-thick aluminum and steel sheets. Fig. 5 shows the four different stages of the SPR process of three sheets.

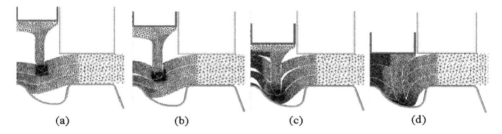

| (a) | (b) | (c) | (d) |

Fig. 5. Four different stages of the SPR of three sheets (Bouchard et al., 2008).

Casalino et al. (Casalino et al., 2008) proposed equations for governing the onset and propagation of crack, the plastic deformation, the space discretization, the time integration, and the contact evolution during the SPR process. A case study of the SPR of two sheets of the 6060T4 aluminum alloy with a steel rivet was performed using the LS-DYNA FE code. Some numerical problems entangled with the model setup and running were resolved and good agreement with experimental results was found in terms of joint cross-sectional shape and force–displacement curve. Fig. 6 shows the initial and final configuration of the joint.

The SPR process currently utilizes high-strength steel rivets. The combination between steel rivets with an aluminum car body not only makes recycling time consuming and costly, but also galvanic corrosion. Galvanic corrosion occurs when dissimilar, conductive materials are joined and the ingress of water forms an electrolytic cell. In this type of corrosion, the material is uniformly corroded as the anodic and cathodic regions moves and reverses from time to time (He et al., 2008). Hoang et al. (Hoang et al., 2010) investigated the possibility of

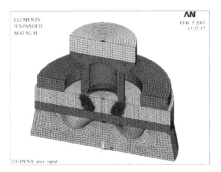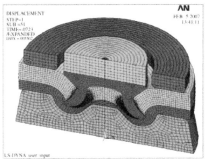

Fig. 6. Initial and final configuration of the joint (Casalino et al., 2008).

replacing steel self-piercing rivets with aluminum ones, when using a conventional die in accordance with the Boellhoff standards. An experimental program was carried out. The test results were exploited in terms of the riveting force–displacement curves and cross-sectional geometries of the riveted joints. The test data were also used to validate a 2D-axisymmetric FEA model. The mechanical behavior of a riveted connection using an aluminum rivet under quasi-static loading conditions was experimentally studied and compared with corresponding tests using a steel rivet.

2.2 Static and fatigue behavior of SPR joints

Sheet material joints are often the structural weakest point of a mechanical system. Consequently a considerable amount of FEAs have been carried out on the static and fatigue behavior of SPR joints. As SPR is considered to be an alternative to spot welding, most research studies have focused on comparisons of the mechanical behavior of joints manufactured by these techniques. Research in this area has shown that SPR gives joints of comparable static strength and superior fatigue behavior to spot welding, whilst also producing promising results in peel and shear testing. Fig. 7 compares the fatigue behavior of three typical joining techniques (Cai et al., 2005).

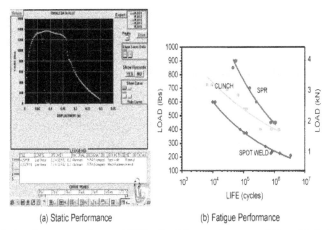

(a) Static Performance   (b) Fatigue Performance

Fig. 7. Fatigue behavior comparison of SPR, clinching and spot welding (Cai et al., 2005).

The fatigue behavior of single and double riveted joints made from aluminum alloy 5754-O has been studied by Iyer et al. (Iyer et al., 2005). A 3D elastic FEA showed that crack initiation occurred at the region of maximum tensile stress. This finding highlights the importance of the cold-formed geometric nonlinearities in determining joint's mechanical strength. The authors also found that both the fatigue and static strength of double-riveted SPR joints was strongly dependent on the orientation (direction that the rivet is inserted; either both from the same side or one from each side) combination of the rivets. The study shows that the analyses were useful for interpreting experimental observations of fatigue crack initiation location, life and fretting damage severity.

Porcaro et al. (Porcaro et al., 2004) developed a numerical model of a riveted structure with the finite element code LS-DYNA to investigate the behavior of a single-riveted joint under combined pull-out and static shear loading conditions. The rivet was represented using the *CONSTRAINED_SPOTWELD card and included failure criteria based on a critical plastic failure strain and the force envelope. Validation was achieved from static and dynamic laboratory test results. The numerical analyses of these components provided a direct check of the accuracy and robustness of the numerical model. The same authors also generated an accurate 3D numerical model of different types of riveted connections, subjected to various loading conditions (Porcaro et al., 2006). An algorithm was generated in order to transfer all the information from the 2D numerical model of the riveting process to the 3D numerical model of the connection. Again the model was validated against the experimental results.

Kim et al. (Kim et al., 2005) tried to evaluate the structural stiffness and fatigue life of SPR joint specimens experimental and numerically by FEA modeling in accordance with the FEMFAT guidelines. The authors found that even though the joint stiffness was independent of substrate thicknesses, the fatigue life was dependant on substrate material and thickness. In research paper of Galtier and Duchet (Galtier & Duchet, 2007), the main parameters that influence the fatigue behavior of sheet material assemblies were presented and some comparisons were made within sheet material joining techniques. It was found that the SPR joint fatigue strength mainly depends on the grade and thickness of the sheet placed on the punch side. In Lim's research paper (Lim, 2008), the simulations of various SPR specimens (coach-peel specimen, cross-tension specimen, tensile-shear specimen, pure-shear specimen) were performed to predict the fatigue life of SPR connections under different shape combinations. FEA models of various SPR specimens were developed using a FEMFAT SPOT SPR pre-processor.

The SPR process has the disadvantage of needing high setting forces typically around 10 times those used for spot-welding. This large setting force can cause severe joint distortion and this will affect the assembly dimensions. SPR joint distortion was found to be much larger, about two to four times the magnitude, than that from resistance spot welding. It was suggested that the inclusion of SPR joint distortion is generally needed for accurate global assembly predictions. To include this localized SPR effect, Sui et al. (Sui et al., 2007) has built a FE model for simulating SPR process of 1.15 mm AA6016T4+1.5 mm AA5182O sheets. The results show that punching load was significantly affected by the deformation of rivet shank and the distortion of the joints was mainly affected by the binder force and the blankholder diameter.

The structural behavior of the SPR joints under static and dynamic loading conditions and how they are modeled in large-scale crash analyses are crucial to the design of the overall structure. Therefore, there is a need to perform dynamic testing on elementary joints in order to study its dynamic behavior. Porcaro et al. (Porcaro et al., 2008) investigated the SPR connections under quasi-static and dynamic loading conditions. Two new specimen geometries with a single rivet were designed in order to study the riveted connections under pull-out and shear impact loading conditions using a viscoelastic split Hopkinson pressure bar. 3D numerical simulations of the SPR connections were performed using the explicit finite-element code LS-DYNA. Static and dynamic tests were simulated using a simplified model that included only the specimen and the clamping blocks that connected the specimen to the bars.

2.3 Vibration behavior of SPR joints

Despite these impressive developments, unfortunately, research in the area of dynamic properties of the SPR joints is relatively unexplored. Hence there is a considerable need for contribution of knowledge in the understanding of the vibration characteristics of SPR joints. Research work by He et al. (He et al., 2006, 2007, Dong et al., 2010) investigated in detail the free vibration characteristics of single lap-jointed SPR beam. The focus of the analysis was to reveal the influence on the natural frequencies and mode shapes of the characteristics of the substrates. These investigations were performed by means of the 3D FEA. In order to obtain the sophisticated features such as design optimization, ANSYS Parametric Design Language (APDL) was used in the analysis. The natural frequencies (eigenvalues) and mode shapes (eigenvectors) of the free vibration of these beams were calculated for different combinations of the substrates' Young's modulus and Poisson's ratio.

By means of a parametric analysis, the influence of the Young's modulus and Poisson's ratio of the lightweight sheet materials on the natural frequencies, natural frequency ratios and mode shapes of the single lap-jointed SPR beams is deduced. Numerical examples show that the natural frequencies increase significantly as the Young's modulus of the substrates increases, but very little change is encountered for change in the substrates' Poisson's ratio. Fig. 8 shows effects of mechanical properties of sheets on torsional natural frequency. It is clear that the torsional natural frequencies increase as the Young's modulus of the sheet increases.

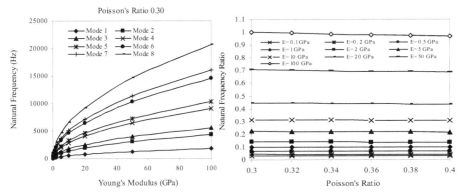

Fig. 8. Torsional natural frequencies versus Young's modulus of sheets for vs=0.30 (He et al., 2006).

Fig. 9 shows the first six mode shapes of the single lap-jointed encastre SPR beam. The mode shapes show that there are different deformations in the jointed section of SPR beams. These different deformations may cause different natural frequency values and different stress distributions. This data will enable appropriate choice of the mechanical properties of the substrates, especially Young's modulus, in order to achieve and maintain a satisfactory level of both static and dynamic integrity of the SPR structures.

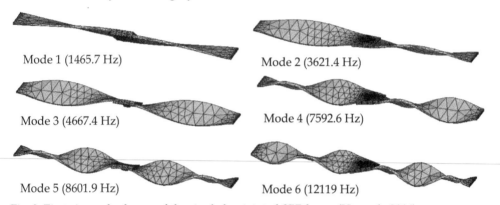

Mode 1 (1465.7 Hz)   Mode 2 (3621.4 Hz)

Mode 3 (4667.4 Hz)   Mode 4 (7592.6 Hz)

Mode 5 (8601.9 Hz)   Mode 6 (12119 Hz)

Fig. 9. First six mode shapes of the single lap-jointed SPR beam (He et al., 2006).

3. FEA of mechanical clinching

The mechanical clinching process is a method of joining sheet metal or extrusions by localized cold forming of materials. The result is an interlocking friction joint between two or more layers of material formed by a punch into a special die. Depending on the tooling sets used, clinched joints can be made with or without the need for cutting. By using a round tool type, materials are only deformed. If a square tool is used, however, both deformation and cutting of materials are required. The principle of clinching with a round tool is given in Fig. 10. Fig. 11 shows the clinching machine and clinching tools in Innovative Manufacturing Research Centre of Kunming University of Science and Technology.

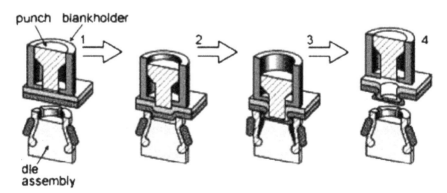

Fig. 10. Principle of clinching with a round tool.

Fig. 11. Clinching machine and clinching tools.

3.1 Clinching process

To fully understand the behavior of clinched joints, the FE model must include all the information from the clinching process. The information that can be obtained from the clinching process simulation includes: metal flow and details of die fill, distribution of strains, strain rate and stresses in the material, distribution of pressure at die–material interface, and the influence of properties like friction. The resolution of such problems is confronted with numerous nonlinear problems such as large deformations, material plasticity, and contact interactions. Several numerical techniques can be used for the simulation of such problems (dynamic or static implicit and explicit methods) and different industrial software (ABAQUS, ADINA, LS-DYNA, and MARC) already allows the simulation of the clinch forming process.

Feng et al. (Feng et al., 2011) presented a LS-DYNA 2D axi-symmetric FE model to predict the magnitude and distribution of deformation associated with the clinching process. The flow stress of the work-material was taken as a function of strain and strain rate. The shape of the clinch joint and the stress, strain, and damage in substrates were determined. Fig. 12 shows the FE simulation of the clinching process and Fig. 13 shows the cross section comparison between simulation and test.

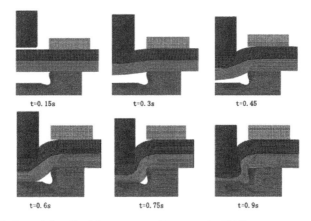

Fig. 12. FE simulation of the clinching process (Feng et al., 2011).

Fig. 13. Cross-section comparison between simulation and test (Feng et al., 2011).

A finite element procedure with automatic remeshing technique has been developed by Hamel et al. (Hamel et al., 2000) using ABAQUS FE software to specifically simulate the clinching process. The resolution of the updated Lagrangian formulation is based on a static explicit approach. The integration of the elastic–plastic behavior law is realized with a Simo and Taylor algorithm, and the contact conditions are insured by a penalty method. The results are compared with experimental data and numerical results calculated with a static implicit method as shown in Fig. 14.

A new clinching process, namely flat clinching, has been introduced by Borsellino et al. (Borsellino et al., 2007). After a press clinching process, the joined sheets have been deformed by a punch with a lower diameter against a flat die. In this way, a new configuration is created with a geometry that has no discontinuity on the bottom surface. Tensile tests have been done to compare the joints strength among the various clinching processes. A FEA has been performed to optimize the process.

Neugebauer et al. (Neugebauer et al., 2008) presented another new clinching method, namely dieless clinching, that works with a flat anvil as a counter tool, thus offering important benefits for the joining of magnesium. Dieless clinching allows mechanical joining of magnesium materials with very short process times because components are heated in less than 3 s. Deformation simulations with DEFORM were used to study the impact that modified punch geometry parameters have on dieless-clinched connections of various combinations of materials and component thicknesses.

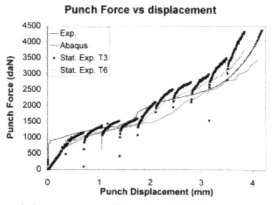

Fig. 14. Computed punch force variation (Hamel et al., 2000).

The information obtained from the process can not only lead to an improvement in die and process design achieving reduction in cost and improvement in the quality of the products but also be used to set initial parameters for a numerical model of the clinched joints used in further mechanical property studies such as static and fatigue analysis, crash analysis, and assembly dimensional prediction etc.

3.2 Strength of clinched joints

The strength of the clinching has been compared with other joining techniques, such as self-pierce riveting and spot welding by researchers (Cai et al., 2005, Lennon et al., 1999). Although the static strength of clinched joints is lower than that of other joints, but the fatigue strength of clinched joints is comparable to that of other joints, and the strength of the clinched joints is more consistent with a significantly lower variation across a range of samples.

Varis and Lepistö (Varis & Lepistö, 2003) proposed a procedure to select an appropriate combination of clinching tools, so that the maximum load under shearing test could be obtained. The calculations considered the final bottom thickness of the joint and the height of the bent area, for each of the analyzed tool combinations. FEAs were performed in order to verify the procedure, although both methods can be used either separately or together to establish optimal clinching parameters.

Carboni et al.'s work (Carboni et al., 2006) focused on a deeper study of the mechanical behavior of clinching in terms of static, fatigue, and residual strength tests after fatigue damage. Fractographic observations showed three different failure modes whose occurrence depends on the maximum applied load and on the stress ratio. Results were supported by FEA showing that the failure regions of the clinched joints correspond to those with high stress concentrations as shown in Fig. 15.

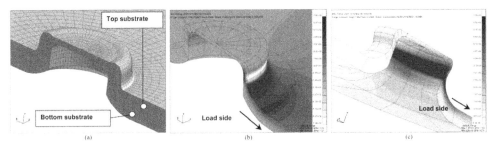

Fig. 15. FEA of clinched joints (Carboni et al., 2006): (a) brick modelling of an indentation point, (b) max principal stress in the bottom substrate and (c) maximum Von Mises stress around the neck of the indentation point.

The FEA of the clinch joining of metallic sheets has been carried out by de Paula et al. (de Paula et al., 2007). The simulations covered the effect of these changes on the joint undercut and neck thickness. The relevant geometrical aspects of the punch/die set were determined, and the importance of an adequate undercut on the joint strength was confirmed.

Clinching tools geometry optimization has been dealt systematically by Oudjene et al. (Oudjene et al., 2008). A parametrical study, based on the Taguchi's method, has been conducted to properly study the effects of tools geometry on the clinch joint resistance as well as on its shape. The separation of the sheets is simulated using ABAQUS/Explicit in order to evaluate the resistance of clinch joints. In a similar study (Oudjene et al., 2009), a response surface methodology, based on Moving Least-Square approximation and adaptive moving region of interest, is presented for shape optimization of clinching tools. The geometries of both the punch and the die are optimized to improve the joints resistance to tensile loading. Fig. 16 shows the FEA of equivalent plastic strain distribution in a clinched joint.

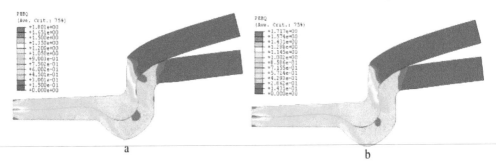

a   b

Fig. 16. Equivalent plastic strain distribution (Oudjene et al., 2009): (a) without remeshing and (b) with remeshing.

3.3 Vibration behavior of clinched joints

Mechanical structures assembled by mechanical clinching are expected to possess a high damping capacity. However, few investigations have been carried out for clarifying the damping characteristics of clinching structures and to establish an estimation method for the damping capacity.

Research papers by He et al. (He et al., 2009a, Gao et al., 2010, Zhang et al., 2010) investigated in detail the free vibration characteristics of single lap-joint clinched joints. Fig. 17 shows the first eight transverse mode shapes of the single lap-joint encastre clinched joint corresponding to material Poisson's ratio 0.33, Young's modulus 70 GPa, and density 2700 kg/m³ (He et al., 2009a).

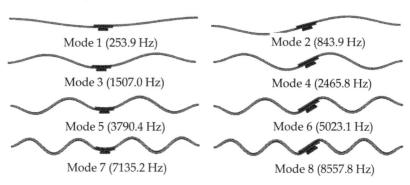

Fig. 17. First eight transverse mode shapes of the single lap-joint encastre clinched joint (He et al., 2009a).

It can be seen from Fig. 17 that the amplitudes of vibration at the midlength of the joints are different for the odd and even modes. For the odd modes (1, 3, 5, and 7), symmetry is seen about the midlength position. At these positions, the amplitudes of transverse free vibration are about equal to the peak amplitude. Thus, the geometry of the lap joint is very important and has a very significant effect on the dynamic response of the lap-jointed encastre clinched joints. Conversely, for the even modes 2, 4, 6, and 8, antisymmetry is seen about the midlength position, and the amplitude of transverse free vibration at this position is approximately zero. Hence, variations in the structure of the lap joint have relatively less effect on the dynamic response of the lap-jointed encastre clinched joints.

4. FEA of structural adhesive bonding

Adhesive bonding has come to be widely used in different industrial fields with the recent development of tough structural adhesives and the substantial improvement in the strength of adhesive joints. Up until 2009, for example, the market demand for automobile adhesives was viewed as increasing very fast and the average per-vehicle consumption of adhesives/sealants was around 20 kg. The structural automotive adhesives would have an average annual growth rate of greater than 7% over the next five years. In the aerospace industry, more and more adhesives have been used in the construction of the aircraft culminating in the Boeing 787 and the Airbus A350 both of which contain more than 50% bonded structure (Speth et al., 2010).

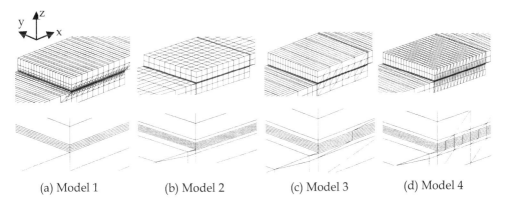

(a) Model 1   (b) Model 2   (c) Model 3   (d) Model 4

Fig. 18. Four examples of smooth transition between the adherends and adhesive (He, 2011c).

The FEA has the great advantage and is frequently used in adhesive bonding since 1970s. In the case of FEA of adhesively bonded joints, however, the thickness of adhesive layer is much smaller than that of the adherends. The finite element mesh must accommodate both the small dimension of the adhesive layer and the larger dimension of the remainder of the whole model. Moreover, the failures of adhesively bonded joints usually occur inside the adhesive layer. It is essential to model the adhesive layer by a finite element mesh which is smaller than the adhesive layer thickness. The result is that the finite element mesh must be several orders of magnitude more refined in a very small region than is needed in the rest of the joint. Thus the number of degrees of freedom in an adhesively bonded joint is rather high. It is also important that a smooth transition between the adherends and adhesive be provided.

Some finite element models have been created by He (He, 2011c) for analyzing the behaviour of bonded joints. Fig. 18 shows four finite element models of bonded joint. From the results of the comparison of the finite element models, it is clear that within the 4 models presented here, model 3 has a moderate size of elements and nodes. Meanwhile, smooth transition between the adherends and the adhesive was also obtained in both x and z directions. Model 3 is then expected to be the best of the 4 models.

4.1 Static loading analysis of adhesive bonded joints

Adhesively bonded joints occurring in practice are designed to carry a given set of loads. The subsequent loads on the adhesive are then a function of the geometry of the joint. A common type of mechanical loading encountered by adhesively bonded joints such as in civil engineering is static loading. In addition, static analysis of adhesively bonded joints will provide a basis for further fatigue, dynamic analyses of the joints.

4.1.1 Stress distribution

The adhesively bonded joints should be designed to minimize stress concentrations. Some stresses, such as peel and cleavage, should also be minimized since these stresses are ultimately responsible for the failure of the joints. In order to determine the physical nature of stress distribution in adhesively bonded joints, the single-lap joints (SLJs) have been investigated by many researchers owing to its simple and convenient test geometry. The lap-joint problem is three-dimensional although it has a simple geometry. The stress behavior of the SLJs is rather complex since bending is induced during the deformation. It is found that the highest stresses and strain in the SLJs occur in regions at the edge of the overlap. The use of the FEA enables the distributions in the critical regions to be predicted with reasonable accuracy.

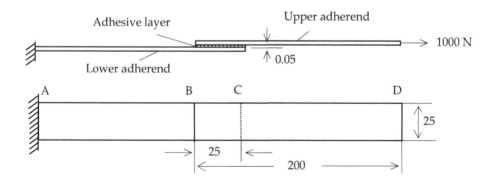

Fig. 19. A bonded joint with rubbery adhesive layer (He, 2011a).

The stress distribution in adhesively bonded joints with rubbery adhesives has been studied by He (He, 2011a). The 3-D FEA software was used to model the joint and predict the stress distribution along the whole joint. Fig. 19 shows a bonded joint with rubbery adhesive layer and Fig. 20 shows the distributions of 6 components of stresses in the lap section. The FEA results indicated that there are stress discontinuities existing in the stress distribution within the adhesive layer and adherends at the lower interface and the upper interface of the bonded section for most of the stress components. The FEA results also show that the stress field in the whole joint is dominated by the normal stresses components S_{11}, S_{33} and the shear stress component S_{13}.

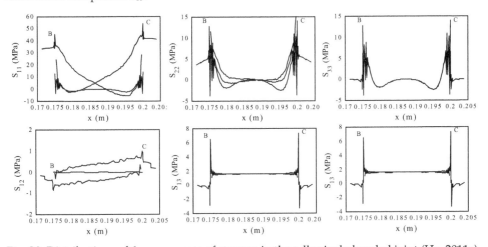

Fig. 20. Distributions of 6 components of stresses in the adhesively bonded joint (He, 2011a).

4.1.2 Stress singularity

Differences in mechanical properties between adherents and adhesive may cause stress singularity at the free edge of adhesively bonded joints. The stress singularity leads to the failure of the bonding part in joints. It is very important to analyze a stress singularity field for evaluating the strength of adhesively bonded joints. Although FEA is well suited to model almost any geometrical shape, traditional finite elements are incapable of correctly resolving the stress state at junctions of dissimilar materials because of the unbounded nature of the stresses. To avoid any adverse effects from the singularity point alternative approaches need to be sought.

Kilic et al. (Kilic et al., 2006) presented a finite element technique utilizing a global (special) element coupled with traditional elements. The global element includes the singular behavior at the junction of dissimilar materials with or without traction-free surfaces. Goglio and Rossetto (Goglio & Rossetto, 2010) explored recently the effects of the main geometrical features of an adhesive SLJ (subjected to tensile stress) on the singular stress field near to the interface end. Firstly an analysis on a bi-material block was carried out to evaluate the accuracy obtainable from FEA by comparison with the analytical solution for the singularity given by the Bogy determinant. Then the study on the SLJs was carried out by varying both macroscopic (bond length and thickness) and local (edge shape and angle) parameters for a total of 30 cases. It was confirmed that the angle play an important party in reducing the singular stresses. Fig. 21 shows the finite element mesh of the joint.

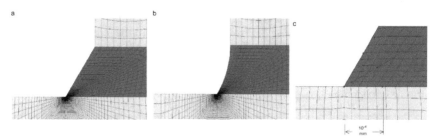

Fig. 21. Finite element mesh of the joint (Goglio & Rossetto, 2010): (a) straight edge; (b) fillet edge; (c) detail view of the elements near to the corner, representative of both cases (a) and (b).

4.1.3 Damage modeling

Damage modeling approach is being increasingly used to simulate fracture and debonding processes in adhesively bonded joints. The techniques for damage modeling can be divided into either local or continuum approaches. In the continuum approach the damage is modeled over a finite region. The local approach, where the damage is confined to zero volume lines and surfaces in 2-D and 3-D, respectively, is often referred to as cohesive zone approach.

Martiny et al. (Martiny et al., 2008) carried out numerical simulations of the steady-state fracture of adhesively bonded joints in various peel test configurations. The model was based on a multiscale approach involving the simulation of the continuum elasto-plastic response of the adherends and the adhesive layer, as well as of the fracture process taking place inside the adhesive layer using a cohesive zone formulation.

4.2 Environmental and fatigue behavior of adhesive bonded joints

Structural adhesives are generally thermosets such as acrylic, epoxy, polyurethane and phenolic adhesives. They will be affected by environmental conditions and exhibit time dependent characteristics. The lifetime of adhesive joints are difficult to model accurately and their long-term performance cannot easily and reliably be predicted, especially under the combined effects of an aggressive environment and fatigue loading (He, 2011b).

4.2.1 Moisture effects on adhesively bonded joints

The adhesives absorb moisture more than most substrate materials and expand more because of the moisture. Water may affect both the chemical and physical characteristics of adhesives and also the nature of the interfaces between adhesive and adherends.

Hua et al. (Hua et al., 2008) proposed a progressive cohesive failure model to predict the residual strength of adhesively bonded joints using a moisture-dependent critical equivalent plastic strain for the adhesive. A single, moisture-dependent failure parameter, the critical strain, was calibrated using an aged, mixed-mode flexure (MMF) test. The FEA package ABAQUS was used to implement the coupled mechanical-diffusion analyses required. This approach has been extended to butt joints bonded with epoxy adhesive. This involves not only a different adhesive and joint configuration but the high hydro- static stress requires a more realistic yielding model (Hua et al., 2007).

4.2.2 Temperature effects on adhesively bonded joints

A detailed series of experiments and FEA were carried out by Grant et al. (Grant et al., 2009a) to assess the effects of temperature that an automotive joint might experience in service. Tests were carried out at -40 and +90 ⁰C. It was shown that the failure criterion proposed at room temperature is still validat low and high temperatures, the failure envelope moving up and down as the temperature increases or decreases, respectively.

Apalak and Gunes (Apalak & Gunes, 2006) investigated 3D thermal residual stresses occurring in an adhesively bonded functionally graded SLJ subjected to a uniform cooling. They concluded that the free edges of adhesive–adherend interfaces and the corresponding adherend regions are the most critical regions, and the adherend edge conditions play more important role in the critical adherend and adhesive stresses.

4.2.3 Fatigue damage modeling

For adhesives, the presence of fatigue loading is found to lead to a much lower resistance to crack growth than under monotonic loading. The fatigue behavior of adhesively bonded joints needs a significant research improvement in order to understand the failure mechanisms and the influence of parameters such as surface pre-treatment, adhesive thickness or adherends thickness.

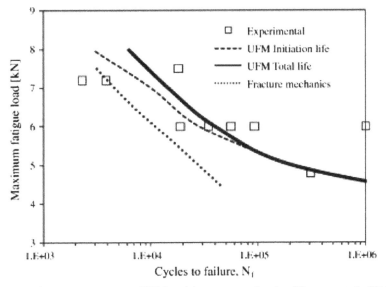

Fig. 22. Extended L–N curves using UFM and fracture mechanics (Shenoy et al., 2010).

A procedure to predict fatigue crack growth in adhesively bonded joints was developed by Pirondi and Moroni (Pirondi & Moroni, 2010) within the framework of Cohesive Zone Model (CZM) and FEA. The idea is to link the fatigue damage rate in the cohesive elements to the macroscopic crack growth rate through a damage homogenization criterion. In Shenoy et al.'s study (Shenoy et al., 2010), a unified fatigue methodology (UFM) was proposed to predict the fatigue behavior of adhesively bonded joints. In this methodology a damage evolution law is used to predict the main parameters governing fatigue life. The model is able to predict the damage evolution, crack initiation and propagation lives, strength and stiffness degradation and the backface strain (BFS) during fatigue loading. The model is able to unify previous approaches based on total life, strength or stiffness wearout, BFS monitoring and crack initiation and propagation modeling. Fig. 22 shows the extended L-N curve using UFM and fracture mechanics. It can be seen that the UFM approach, which accounts for both initiation and propagation can provide a good prediction of the total fatigue life at all loads.

4.3 Dynamic loading analysis of adhesive bonded joints

Adhesive bonding offers advantages on acoustic isolation and vibration attenuation relatively to other conventional joining processes. Mechanical structures assembled by adhesively bonding are expected to possess a high damping capacity because of the high damping capacity of the adhesives.

4.3.1 Structural damping

Investigations have been carried out for clarifying the damping characteristics of adhesively bonded structures and to establish an estimation method for the damping capacity. Research work by He (He, 2010a) studied the influence of adhesive layer thickness on the dynamic behavior of the single-lap adhesive joints. The results showed that the composite damping of the single-lap adhesive joint increases as the thickness of the adhesive layer increases.

In a research paper by Apalak and Yildirim (Apalak & Yildirim, 2007), the 3D transient vibration attenuation of an adhesively bonded cantilevered SLJ was controlled using actuators. The transient variation of the control force was expressed by a periodic function so that the damped vibration of the SLJ was decreased. Optimal transient control force history and optimal actuator position were determined using the Open Loop Control Approach (OLCA) and Genetic Algorithm.

4.3.2 Modes of vibration

With the increase in the use of adhesively bonded joints in primary structures, such as aircraft and automotive structures, reliable and cost-effective techniques for structural health monitoring (SHM) of adhesive bonding are needed. Modal and vibration-based analysis, when combined with validated FEA, can provide a key tool for SHM of adhesive bonding.

Gunes et al. (Gunes et al., 2010) investigated the free vibration behavior of an adhesively bonded functionally graded SLJ, which composed of ceramic (Al2O3) and metal (Ni) phases varying through the plate thickness. The effects of the similar and dissimilar material composition variations through-the-thicknesses of both upper and lower plates on the natural frequencies and corresponding mode shapes of the adhesive joint were investigated using both the FEA and the back-propagation artificial neural network (ANN) method. A series of the free vibration analyses were carried out for various random values of the geometrical parameters and the through-the-thickness material composition so that a suitable ANN model could be trained successfully.

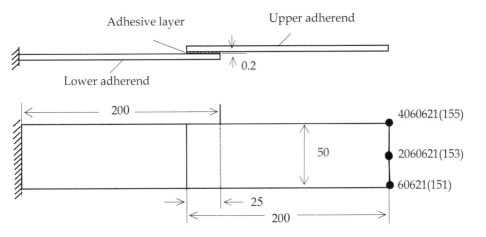

Fig. 23. Location of nodes at the free edge of the beam (He, 2009).

The ABAQUS FEA software was used by He (He, 2009) to predict the natural frequencies, mode shapes and frequency response functions (FRFs) of adhesively lap-jointed beams. In the case of forced vibration of the single lap-jointed cantilevered beam, some typical points on the free edge were chosen for response points because they can better represent the dynamical characteristic of the beams. Nodes 151, 153 and 155 in Fig. 23 are the nodes at the two corners and center of the free edge of the beam (the corresponding nodes in the FE mesh are 60621, 2060621 and 4060621, respectively).

The overlay of the FRFs predicted using FEA and measured experimentally at the 2 corners and centre of the free edge of the beam are shown in Fig. 24. It can be seen that the measued FRFs are close to the predicted FRFs for the first few modes of vibration of the beam. For the higher order modes of vibration, there is considerable discrepancy between the measured and predicted FRFs. This discrepancy can be attributed to locations and additional masses of force transducer and accelerometer. In order to fully excite the beams, the force transducer was connected to the location which was 20% of the length of the beam from the clamped end and very close to a free edge. The accelerometer was connected to different locations of the beam for obtaining precise mode shapes. As a result, the natural frequencies from experiments are lower than those predicted using FEA and some complexity appear in the FRFs graphics.

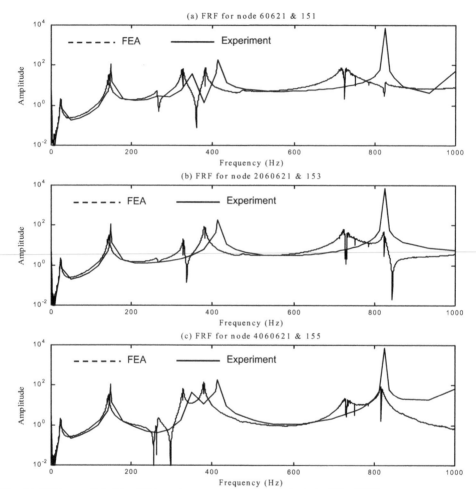

Fig. 24. FRFs predicted by FEA and measured using the test rig (He, 2009).

5. FEA of hybrid joints

It is also important for one joining process to benefit from the advantages of other joining techniques. These can be done by combining one joining process with other joining techniques and are referred to as hybrid joining processes. A number of researchers have carried out mechanical performances of the hybrid joints in different materials with various load conditions. Their study shows that the combination produced a much stronger joint in both static and fatigue tests.

5.1 Clinch-bonded hybrid joints

Pirondi and Moroni (Pirondi & Moroni, 2009) simulated the failure behavior of clinch-bonded and rivet-bonded hybrid joints using the FEA. The analyses were conducted using

Arbitrary Lagrangian–Eulerian (ALE) adaptive meshing to avoid excessive element distortion and mass scaling to increase the minimum time increment in explicit analyses. The authors concluded that different damage models, tuned with experiments performed on simple joints (riveted, clinched or adhesively bonded), can be combined in a unique model to simulate effectively the failure behavior of hybrid joints. A detailed series of tests and FEA were conducted by Grant et al. (Grant et al., 2009b) for clinch-bonded hybrid joints. The experimental results were compared with spot welded joints and adhesively bonded double-lap joints. It was concluded that this joint fails because of large plastic deformation in the adherend. By means of 3D ANSYS FEA, the influence of Young's modulus and Poisson's ratio of the structural adhesives on the natural frequencies, natural frequency ratios, and mode shapes of the single-lap clinch-bonded hybrid joints was deduced by He et al. (He et al., 2010b).

5.2 SPR-bonded hybrid joints

The torsional free vibration behavior of SLJ encastre hybrid SPR-bonded beams has been studied by He et al. (He et al., 2009b) using the commercially available ANSYS FEA program. The mode shapes showed that there are different deformations in the jointed section of the odd and even modes. These different deformations may result different dynamic response and different stress distributions. Another research work by the same authors investigated in detail the free transverse vibration characteristics of single-lap SPR-bonded hybrid joints (He et al., 2010a). The FEA results showed that the stiffer adhesive is likely to suffer fatigue failure and debonding more often than the softer adhesive. These deformations may result in relatively high stresses in the adhesive layers and initiate local cracking and delamination failures.

6. Conclusions

Some joining techniques have become increasingly popular alternatives to traditional spot welding due to the growing use of alternative materials which are difficult or impossible to weld. Adequate understanding of the behavior of joints is necessary to ensure efficiency, safety and reliability of such joining structures. However, accurate and reliable modeling of joining structures is still a difficult task as the mechanical behavior of these joints is not only influenced by the geometric characteristics of the joint but also by different factors and their combinations.

The information that can be obtained from the FEA of sheet material joints includes: differences in the basic mechanical properties, hygrothermal behavior, occurrences of high stress gradients in certain regions of the joints. An accurate FEA model of sheet material joint must be able to predict failure in the substrates and at the interfaces, and must also account for full non-linear material behavior.

In this chapter the research and progress in FEA of sheet material joints are critically reviewed and current trends in the application of FEA are mentioned. It is concluded that the FEA of sheet material joints will help future applications of sheet material joining by allowing different parameters to be selected to give as large a process window as possible for joint manufacture. This will allow many tests to be simulated that would currently take too long to perform or be prohibitively expensive in practice, such as modifications to geometry or material properties.

7. Acknowledgment

This study is partially supported by National Science Foundation of China (Grant No. 50965009).

8. References

Abe, Y.; Kato, T. & Mori, K. (2006). Joinability of aluminum alloy and mild steel sheets by self piercing rivet. *Journal of Materials Processing Technology*, Vol. 177, No. 1-3, (July 2006), pp. 417-421, ISSN 0924-0136

Abe, Y.; Kato, T. & Mori, K. (2009). Self-piercing riveting of high tensile strength steel and aluminum alloy sheets using conventional rivet and die. *Journal of Materials Processing Technology*, Vol. 209, No. 8, (April 2009), pp. 3914-3922, ISSN 0924-0136

Apalak, M. K. &; Gunes, R. (2006). Thermal residual stresses in an adhesively-bonded functionally graded single-lap joint. *Journal of Adhesion Science and Technology*, Vol. 20, No. 12, (December 2006), pp. 1295-1320, ISSN 0169-4243

Apalak, M. K. & Yildirim, M. (2007). Optimal vibration attenuation of an adhesively-bonded cantilevered single-lap joint. *Journal of Adhesion Science and Technology*, Vol. 21, No. 3-4, (April 2007), pp. 267-286, ISSN 0169-4243

Atzeni, E.; Ippolito, R. & Settineri, L. (2007). FEM modeling of self-piercing riveted joint. *Key Engineering Materials*, Vol. 344, (July 2007), pp. 655–662, ISSN 1013-9826

Atzeni, E.; Ippolito, R. & Settineri, L. (2009). Experimental and numerical appraisal of self-piercing riveting. *CIRP Annals - Manufacturing Technology*, Vol. 58, No. 1, (December 2009), pp. 17-20, ISSN 0007-8506

Borsellino, C.; Di Bella, G. & Ruisi, V.F. (2007). Study of new joining technique: flat clinching. *Key Engineering Materials*, Vol. 344, (July 2007), pp. 685–692, ISSN 1013-9826

Bouchard, P.O.; Laurent, T. & Tollier, L. (2008). Numerical modeling of self-pierce riveting — from riveting process modeling down to structural analysis. *Journal of Materials Processing Technology*, Vol. 202, No. 1-3, (June 2008), pp. 290–300, ISSN 0924-0136

Cai, W.; Wang, P. & Yang, W. (2005). Assembly dimensional prediction for self-piercing riveted aluminium panels. *International Journal of Machine Tools and Manufacture*, Vols. 45, No. 6, (May 2005), pp. 695-704, ISSN 0890-6955

Carboni, M.; Beretta, S. & Monno, M. (2006). Fatigue behavior of tensile-shear loaded clinched joints. *Engineering Fracture Mechanics*, Vol. 73, No. 2, (January 2006), pp. 178-190, ISSN 0013-7944

Casalino, G.; Rotondo, A. & Ludovico, A. (2008). On the numerical modeling of the multiphysics self piercing riveting process based on the finite element technique. *Advanced Engineering Software*, Vol. 39, No. 9, (September 2008), pp. 787-795, ISSN 0965-9978

de Paula, A.A.; Aguilar, M.T.P.; Pertence, A.E.M. & Cetlin, P.R. (2007). Finite element simulations of the clinch joining of metallic sheets. *Journal of Materials Processing Technology*, Vol. 182, No. 1-3, (February 2007), pp. 352–357, ISSN 0924-0136

Dong, B. ; He, X. & Zhang, W. (2010). Harmonic response analysis of single lap self-piercing riveted joints. *New Technology & New Process (in Chinese)*, Vol. 275, (November 2010), pp. 51-54, ISSN 1003-5311

Feng, M. ; He, X. ; Yan K. & Zhang, Y. (2011). Numerical simulation and analysis of clinch process. *Science Technology and Engineering (in Chinese)*, Vol. 11, No. 23, (August 2011), pp. 5538-5541, ISSN 1671-1815

Galtier, A. & Duchet, M. (2007). Fatigue behaviour of high strength steel thin sheet assemblies. *Welding in the World*, Vol. 51, No. 3-4, (March/April 2007), pp. 19-27, ISSN 0043-2288

Gao, S. & He, X. (2010). Influence of adhesive characteristics on transverse free vibration of single lap clinched joints. *Machinery (in Chinese)*, Vol. 48, No. 553, (September 2010), pp. 32-34, ISSN 1000-4998

Goglio, L. & Rossetto, M. (2010). Stress intensity factor in bonded joints: Influence of the geometry, *International Journal of Adhesion and Adhesives*, Vol. 30, No. 5, (July 2010), pp. 313-321, ISSN 0143-7496

Grant, L.D.R.; Adams, R.D. & da Silva, L.F.M. (2009a). Effect of the temperature on the strength of adhesively bonded single lap and T joints for the automotive industry. *International Journal of Adhesion and Adhesives*, Vol. 29, No. 5, (July 2009), pp. 535-542, ISSN 0143-7496

Grant, L.D.R.; Adams, R.D. & da Silva, L.F.M. (2009b). Experimental and numerical analysis of clinch (hemflange) joints used in the automotive industry. *Journal of Adhesion Science and Technology*, Vol. 23, No. 12, (September 2009), pp. 1673-1688, ISSN 0169-4243

Gunes, R.; Apalak, M.K. Yildirim, M. & Ozkes, I. (2010). Free vibration analysis of adhesively bonded single lap joints with wide and narrow functionally graded plates. *Composite Structures*, Vol. 92, No. 1, (January 2010), pp. 1-17, ISSN 0263-8223

Hamel, V.; Roelandt, J.M.; Gacel, J.N. & Schmit, F. (2000). Finite element modeling of clinch forming with automatic remeshing. *Computers and Structures*, Vol. 77, No. 2, (June 2000), pp. 185-200, ISSN 0045-7949

He, X.; Pearson, I. & Young, K. (2006). Free vibration characteristics of SPR joints. *Proceedings of 8th Biennial ASME Conference on Engineering Systems Design and Analysis*, pp. 1-7, ISBN_10 0791837793, Torino, Italy, July 4-7, 2006

He, X.; Pearson, I. & Young, K. (2007). Three dimensional finite element analysis of transverse free vibration of self-pierce riveting beam. *Key Engineering Materials*, Vol. 344, (July 2007), pp. 647–654, ISSN 1013-9826

He, X.; Pearson, I. & Young, K. (2008). Self-pierce riveting for sheet materials: State of the art. *Journal of Materials Processing Technology*, Vol.199, No.1-3, (April 2008), pp. 27-36, ISSN 0924-0136

He, X. (2009). Dynamic behaviour of single lap-jointed cantilevered beams. *Key Engineering Materials*, Vols. 413-414, (August 2009), pp. 733-740, ISSN 1013-9826

He, X.; Zhang, W.; Dong, B. Zhu, X. & Gao, S. (2009a). Free vibration characteristics of clinched joints. *Lecture Notes in Engineering and Computer Science*, Vol. 1, (July 2009), pp. 1724-1729, ISBN 978-988-18210-1-0

He, X. ; Dong, B. & Zhu, X. (2009b). Free Vibration Characteristics of Hybrid SPR Beams, *AIP Conference Proceedings* Vol. 1233, pp. 678-683, ISBN 978-0-7354-0778-7, Hong Kong-Macau, China, November 30-Desember 3, 2009

He, X. (2010a). Bond Thickness Effects upon Dynamic Behavior in Adhesive Joints. *Advanced Materials Research*, Vols. 97-101 (April 2011), pp. 3920-3923, ISSN 1022-6680

He, X. (2010b). Recent development in finite element analysis of clinched joints. *International Journal of Advanced Manufacturing Technology*, Vol. 48, (May 2010), pp. 607-612, ISSN 0268-3768

He, X.; Zhu, X. & Dong, B. (2010a). Transverse free vibration analysis of hybrid SPR steel joints. Proceedings of MIMT 2010, p 389-394, ISBN-13: 9780791859544, Sanya, China, January 22-24, 2010

He, X.; Gao, S. & Zhang, W. (2010b). Torsional free vibration characteristics of hybrid clinched joints. *Proceedings of ICMTMA 2010*, Vol. 3, pp. 1027-1030, ISBN-13: 9780769539621, Changsha, China, March 13-14, 2010

He, X. (2011a). Stress Distribution of Bonded Joint with Rubbery Adhesives. *Advanced Materials Research*, Vols. 189-193 (April 2011), pp 3427-3430, ISSN 1022-6680

He, X. (2011b). A review of finite element analysis of adhesively bonded joints. *International Journal of Adhesion and Adhesives*, Vol. 31, No. 4, (June 2011), pp. 248-264, ISSN 0143-7496

He, X. (2011c). Comparisons of finite element models of bonded joints. *Applied Mechanics and Materials*, Vols. 66-68, (September 2011), pp. 2192-2197, ISSN 1660-9336

He, X. (2012). Recent development in finite element analysis of self-piercing riveted joints. *International Journal of Advanced Manufacturing Technology*, Vol. 58, (January 2012), pp. 643-649, ISSN 0268-3768

Hoang, N.H.; Porcaro, R.; Langseth, M. & Hanssen, A.G. (2010). Self-piercing riveting connections using aluminum rivets. *International Journal of Solids and Structure*, Vols. 47, No. (3–4), (February 2010), pp. 427–439, ISSN 0020-7683

Hua, Y.; Crocombe, A.D.; Wahab, M.A. & Ashcroft, I.A. (2007). Continuum damage modelling of environmental degradation in joints bonded with E32 epoxy adhesive. *Journal of Adhesion Science and Technology*, Vol. 21, No. 2, (February 2008), pp. 179-195, ISSN 0169-4243

Hua, Y.; Crocombe, A.D.; Wahab, M.A. & Ashcroft, I.A. (2008). Continuum damage modelling of environmental degradation in joints bonded with EA9321 epoxy adhesive. *International Journal of Adhesion and Adhesives*, Vol. 28, No. 6, (September 2008), pp. 302-313, ISSN 0143-7496

Iyer, K.; Hu, S.J.; Brittman, F.L.; Wang, P.C.; Hayden, D.B. & Marin, S.P. (2005). Fatigue of single- And double-rivet self-piercing riveted lap joints. *Fatigue and Fracture of Engineering Materials and Structures*, Vol. 28, No. 11, (November 2005), pp. 997-1007, ISSN 1460-2695

Kato, T.; Abe, Y. & Mori, K. (2007). Finite element simulation of self-piercing riveting of three aluminum alloy sheets. *Key Engineering Materials*, Vols. 340-341, (July 2007), pp. 1461-1466, ISSN 1013-9826

Kilic, B.; Madenci, E. & Ambur, D.R. (2006). Influence of adhesive spew in bonded single-lap joints, *Engineering Fracture Mechanics*, Vol. 73, No. 11, (July 2006), pp. 1472-1490, ISSN 0013-7944

Kim, M.G.; Kim, J.H.; Lee, K.C. & Yi, W. (2005). Assessment for structural stiffness and fatigue life on self-piercing rivet of car bodies. *Key Engineering Materials*, Vols. 297-300, (July 2005), pp. 2519-2524, ISSN 1013-9826

Lennon, R.; Pedreschi, R. & Sinha, B.P. (1999). Comparative study of some mechanical connections in cold formed steel. *Construction and Building Materials*, Vol. 13, No. 3, (April 1999), pp. 109-116, ISSN 0950-0618

Lim, B. K. (2008). Analysis of fatigue life of SPR(Self-Piercing Riveting) jointed various specimens using FEM. *Materials Science Forum,* Vols. 580-582, (July 2008), pp. 617-620, ISSN 0255-5476

Martiny, Ph.; Lani, F.; Kinloch, A.J. & Pardoen, T. (2008). Numerical analysis of the energy contributions in peel tests: A steady-state multilevel finite element approach. *International Journal of Adhesion and Adhesives,* Vol. 28, No. 4-5, (June 2008), pp. 222-236, ISSN 0143-7496

Mori, K.; Kato, T.; Abe, Y. & Ravshanbek, Y. (2006). Plastic joining of ultra high strength steel and aluminum alloy sheets by self piercing rivet. *CIRP Annals - Manufacturing Technology,* Vol. 55, No. 1, (December 2006), pp. 283-286, ISSN 0007-8506

Neugebauer, R.; Kraus, C. & Dietrich, S. (2008) Advances in mechanical joining of magnesium. *CIRP Annals - Manufacturing Technology,* Vol. 57, No. 1, (December 2008), pp. 283-286, ISSN 0007-8506

Oudjene, M. & Ben-Ayed, L. (2008). On the parametrical study of clinch joining of metallic sheets using the Taguchi method. *Engineering Structures,* Vol. 30, No. 6, (June 2008), pp. 352–357, ISSN 0141-0296

Oudjene, M.; Ben-Ayed, L.; Delamézière, A. & Batoz, J.L. (2009). Shape optimization of clinching tools using the response surface methodology with moving least-square approximation. *Journal of Materials Processing Technology,* Vol. 209, No.1, (January 2009), pp. 289–296, ISSN 0924-0136

Pirondi, A. & Moroni, F. (2009). Clinch-bonded and rivet-bonded hybrid joints: Application of damage models for simulation of forming and failure. *Journal of Adhesion Science and Technology,* Vol. 23, No. 10-11, (July 2009), pp. 1547-1574, ISSN 0169-4243

Pirondi, A. & Moroni, F. (2010). A progressive damage model for the prediction of fatigue crack growth in bonded joints. *Journal of Adhesion,* Vol. 86, No. 5-6, (April 2010), pp. 501-521, ISSN 0021-8464

Porcaro, R.; Hanssen, A.G.; Aalberg, A. & Langseth, M. (2004). Joining of aluminum using self-piercing riveting: Testing, modeling and analysis. *International Journal of Crashworthiness,* Vol. 9, No. 2, (May 2004), pp. 141–154, ISSN 1358-8265

Porcaro, R.; Hanssen, A.G.; Langseth, M. & Aalberg, A. (2006). The behaviour of a self-piercing riveted connection under quasi-static loading conditions. *International Journal of Solids and Structure,* Vols. 43, No. 17, (August 2006), pp. 5110-5131, ISSN 0020-7683

Porcaro, R.; Langseth, M.; Hanssen, A.G.; Zhao, H.; Weyer, S. & Hooputra, H. (2008). Crashworthiness of self-piercing riveted connections. *International Journal of Impact Engineering,* Vols. 35, No. 11, (November 2008), pp. 1251-1266, ISSN 0734-743X

Shenoy, V.; Ashcroft, I.A.; Critchlow, G.W. & Crocombe, A.D. (2010). Unified methodology for the prediction of the fatigue behaviour of adhesively bonded joints. *International Journal of Fatigue,* Vol. 32, No. 8, (August 2010), pp. 1278-1288, ISSN 0142-1123

Speth, D.R.; Yang, Y. & Ritter, G.W. (2010). Qualification of adhesives for marine composite-to-steel applications. *International Journal of Adhesion and Adhesives,* Vol. 30, No. 2, (March 2010), pp. 55-62, ISSN 0143-7496

Sui, B.; Du, D.; Chang, B.; Huang, H. & Wang, L. (2007). Simulation and analysis of self-piercing riveting process in aluminum sheets. *Material Science and Technology (in Chinese),* Vol. 15, No. 5, (October 2007), pp. 713-717, ISSN 1005-0299

Varis, J.P. & Lepistö, J. (2003). A simple testing-based procedure and simulation of the clinching process using finite element analysis for establishing clinching parameters. *Thin-Walled Structures,* Vol. 41, No. 8, (August 2003), pp. 691-709, ISSN 0263-8231

Yan, k.; He, X. & Zhang, Y. (2011). Numerical simulation and quality evaluation of single lap self-piercing rivetiong process. *China Manufacturing Information (in Chinese),* Vol. 40, No. 19, (October 2011), pp. 76-78, ISSN 1672-1616

Zhang, W.; He, X. & Dong, B. (2010). Analysis of harmonic response of single lap clinched joints. *Machinery (in Chinese),* Vol. 48, No. 555, (November 2010), pp. 19-21, ISSN 1000-4998

7

Electromagnetic and Thermal Analysis of Permanent Magnet Synchronous Machines

Nicola Bianchi, Massimo Barcaro and Silverio Bolognani
Department of Electrical Engineering, University of Padova
Italy

1. Introduction

The increasing interest to permanent magnet (PM) synchronous machines is due to the high torque density and high efficiency that they may exhibit exploiting modern PMs. The three–phase winding is supplied by a current–controlled voltage source inverter, which imposes sinewave currents synchronous with the PM rotor. Such machines are more and more used in several applications, with power rating ranging from fractions of Watts to some Megawatts.

After a brief introduction on the PM characteristics, this Chapter illustrates the finite element (FE) analysis of the synchronous PM machines. It summarizes the basic concepts dealing with the electromagnetic analysis, and it describes proper analysis strategies in order to predict the PM machine performance.

2. The permanent magnet machines

In synchronous PM machines, the stator is the same of the induction machines. The rotor can assume different topologies, according to how the PM is placed in it. The machines are distinguished in three classes: surface–mounted PM (SPM) machines, inset PM machine, and interior PM (IPM) machine. Fig. 1(a) shows a cross–section of a four–pole 24–slot SPM machine. There are four PMs mounted with alternate polarity on the surface of the rotor. Fig. 1(b) shows a four–pole inset PM machine, characterized by an iron tooth between each couple of adjacent PMs. Fig. 1(c) shows a four–pole IPM machine, whose rotor is characterized by three flux–barriers per pole. The high number of flux–barriers per pole yield a high rotor anisotropy (Honsinger, 1982).

When the IPM machine is characterized by high anisotropy and moderate PM flux, it is often referred to as PM assisted synchronous reluctance (PMASR) machine. The machine exhibits two torque components: the PM torque and the reluctance torque (Levi, 1984) .

Fig. 2 shows the pictures of a rotor with surface–mounted PMs (a), the lamination of an IPM machine with three flux–barriers per pole (b) and an IPM rotor with two flux–barriers per pole (c).

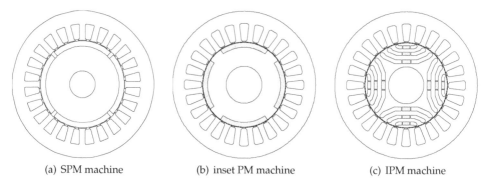

| (a) SPM machine | (b) inset PM machine | (c) IPM machine |

Fig. 1. PM synchronous machines with (a) SPM, (b) inset PM, and (c) IPM rotor.

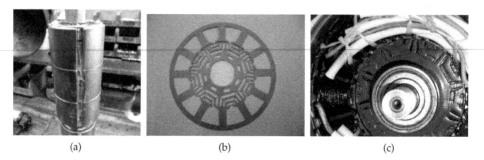

| (a) | (b) | (c) |

Fig. 2. PM machine prototypes: (a) an SPM rotor, (b) and (c) IPM machine laminations.

2.1 Hard magnetic material (permanent magnet)

The permanent magnets are hard magnetic materials (Bozorth, 1993) . They exhibit a very wide magnetic hysteresis loop, as shown in Fig. 3(a). Once they are magnetized and required to sustain a magnetic field, the PMs operate in quadrant II. Both the intrinsic (dashed line) and normal (solid line) hysteresis loops are drawn in Fig. 3(a), as reported in most PM data sheets. The intrinsic curve represents the added magnetic flux density that the PM material produces. The normal curve represents the total magnetic flux density which is carried in combination by the air and by the PM (Coey, 1996) . The recoil line of the demagnetization curve is usually approximated by

$$B_m = B_{rem} + \mu_{rec}\mu_0 H_m \tag{1}$$

where the residual flux density (or remanence) is B_{rem} and the coercive force is H_c (see Fig. 3(a)). The differential relative magnetic permeability of the recoil line is μ_{rec} and is slightly higher than unity.

When the flux density becomes lower than B_{knee}, where the hysteresis curve exhibits a knee in quadrant II, the PM is irreversibly demagnetized and its next operating points move on a locus with lower flux density versus field strength. The minimum flux density in the PMs has to be verified during the magnetic analysis of the machine. Since B_{knee} depends on the temperature, the worst operating condition has to be considered. B_{knee} increases with the temperature, see Fig. 3(b).

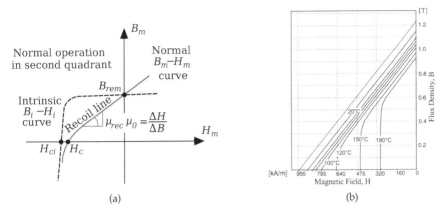

Fig. 3. (a) Characteristic B–H curve of hard magnetic material, (b) example of Neodymium-Iron-Boron PM including the variation of characteristic with the temperature.

	B_{rem} (T)	H_c (kA/m)	Curie T (°C)	T_{max} (°C)	density (kg/m³)	$\Delta B_{rem}/\Delta T$ (%/°C)	$\Delta H_{ci}/\Delta T$ (%/°C)
Ferrite	0.38	250	450	300	4800	-0.20	0.40
SmCo	0.85	570	775	250	8300	-0.04	-0.20
NdFeB	1.15	880	310	180	7450	-0.12	-0.70

Table 1. Main properties of hard magnetic material

In order to avoid the irreversible demagnetization of the PM, it is imperative to verify that the minimum flux density in the PM be always higher than the flux density of the knee B_{knee}.

The key properties of some common PM materials are listed in Table 1. Since the operating temperature has a great impact on the PM characteristic, the rate of change of B_{rem} and of H_c versus temperature is also reported.

3. FE analysis pre–processing

In the most of cases, the the magnetic FE analysis is a two–dimensional (2D) analysis [1]. The three dimensional effects (e.g., leakage inductance of stator end winding, effect of the rotor skewing, etc.) are not considered here. They have to be computed separately, and added to the 2D field solution.

The simulations are carried out by assuming a planar symmetry: the magnetic field is the same in each section of the machine for a z–axis length equal to the machine stack length L_{stk}. The current density $\mathbf{J}$ and the vector magnetic potential $\mathbf{A}$ have only the component normal to the x–y plane, that is, $\mathbf{J} = (0, 0, J_z)$ and $\mathbf{A} = (0, 0, A_z)$. Therefore the magnetic field strength

---

[1] Let us invite the interested reader to download FEMM (Finite Element Method Magnetics) freeware software by David Meeker. Although FEMM exhibits some limitations (i.e., this means that not all field solution problems can be analysed), it allows the most of electrical machine magnetic field computations to be satisfactorily analyzed. It is very easy to be used, it is organized so as to easily understand how the FE method works. For details about the software, please refer to "FEMM User's Manual"

H and flux density **B** vectors have components only in the x–y plane, that is, $\mathbf{H} = (H_x, H_y, 0)$ and $\mathbf{B} = (B_x, B_y, 0)$. Fig. 4(a) shows a cross section of a PM machine in the x–y plane.

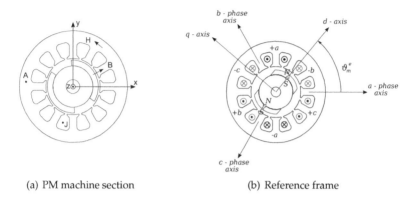

(a) PM machine section   (b) Reference frame

Fig. 4. Electrical machine section in the plane (x, y) and reference frame.

3.1 Reference conventions

Fig. 4(b) shows the two reference frames. The stator reference frame is characterized by the a–, b–, and c–axis, which refer to the axis of the coils of the three phases. The rotor reference frame is characterized by the d– and q–axis, with d–axis aligned with the PM axis.

The relative position between the rotor and the stator reference frames is the mechanical angle ϑ_m, which corresponds to the electrical angle $\vartheta_m^e = p \cdot \vartheta_m$, where p is the number of pole pairs.

3.2 Boundary conditions

In a magnetic field problem, there are four main boundary conditions:
Dirichlet: this condition prescribes a given value of the magnetic vector potential A_z along a line, typically $A_z = 0$. Thus, the flux lines are tangential to this line. This condition is used to confine the field lines into the domain.
As an example, the Dirichlet boundary condition $A_z = 0$ is assigned to the outer periphery of the stator as shown in Fig. 5(a).
Neumann: this condition imposes the flux density lines to be normal to a line. This condition is a default condition in the field problem.
Periodic: this condition is assigned to two lines and imposes that the magnetic vector potential behavior is the same along the two lines, i.e. $A_{z,line1} = A_{z,line2}$.
Anti–periodic: this condition is assigned to two lines and imposes that the magnetic vector potential behavior along one line is opposite to that along the other line, i.e. $A_{z,line1} = -A_{z,line2}$.

3.3 Current sources

In addition to the PMs in the rotor, further sources of magnetic field are the stator currents. They are defined by ideal current generators, connected to stator slot surface. In order to impose a prefixed current to each slot, it is convenient to define one generator per each slot

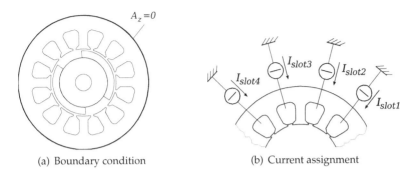

(a) Boundary condition　　　　(b) Current assignment

Fig. 5. Assignment of boundary condition and current sources.

and to assign such a generator to the corresponding slot, as shown in Fig. 5(b).

Each current is defined by its amplitude and sign. Positive sign means that it is along the direction of z–axis (from the sheet towards reader) and negative sign means that it is opposite to the direction of z–axis (from reader towards the sheet). In magnetostatic analysis, only real part of the current is necessary, while in time harmonic analysis (where symbolic phasor notation is adopted) both real and imaginary components have to be assigned.

Alternatively, a current density can be assigned to the various stator slots. Each slot will be characterized by a different current density, according to the total current flowing in the corresponding slot. The peak value of current density in the slot is defined as

$$\hat{J}_{slot} = k_{fill} \sqrt{2} \cdot J_c \tag{2}$$

where k_{fill} is the slot fill factor and J_c is the rms value of the current density in the conductor. For instance, with $k_{fill} = 0.4$, the actual current density $J_c = 6 \text{ A/mm}^2$ corresponds to an equivalent current density $\hat{J}_{slot} = 3.4 \text{ A/mm}^2$ assigned to the stator slot.

3.4 Winding definition

The definition of the winding of the electrical machine is extremely important. Fig. 6(a) shows a two–pole three–phase single–layer winding, characterized by two slots per pole per phase.

The arrangement of the coil sides within the slots is described by means of the slot matrix, whose dimension is $m \times Q$, where m is the number of phases (i.e. $m=3$ in a three–phase machine) and Q is the number of stator slots. The generic element k_{jq} indicates how much the q–th slot is filled by conductors of the j–th phase, where unity means a complete slot fill.

For instance, $k_{jq} = 1$ means that the q–th slot is completely filled by conductors of the j–th phase; $k_{jq} = 0.5$ means that only 50% of the q–th slot is filled by conductors of the j–th phase; and $k_{jq} = 0$ means that no conductor of the j–th phase is in the q–th slot. The element k_{jq} can be either positive or negative sign. The sign refers to the orientation of the coil side.

According to the winding shown in Fig. 6(a), the slot matrix is where the first slot (i.e., $q = 1$) is the slot in the first quadrant closer to the x–axis, and the other slots follow counterclockwise. In order to meet the current Kirchhoff law, the element row sum to zero.

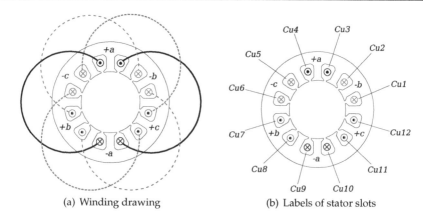

(a) Winding drawing   (b) Labels of stator slots

Fig. 6. Classical representation of the stator winding, and definition of the labels of the stator slots.

q	1	2	3	4	5	6	7	8	9	10	11	12
k_a	0	0	1	1	0	0	0	0	-1	-1	0	0
k_b	-1	-1	0	0	0	0	1	1	0	0	0	0
k_c	0	0	0	0	-1	-1	0	0	0	0	1	1

In order to assign the proper current density in each slot, each of them has been defined by a different label, as sketched in Fig. 6(b). The current in the q–th slot can be expressed as

$$I_{slot,q} = n_{cs}\left(k_{a,q}I_a + k_{b,q}I_b + k_{c,q}I_c\right) \tag{3}$$

where n_{cs} is the number of series conductors per slot, I_a, I_b, and I_c are the currents of the phase a, b, and c, respectively.

4. Post–processing

After the field solution is achieved, various quantities can be computed from the solved structure.

4.1 Flux line plot

The flux lines of the solved structure, shown in Fig. 7(a), give an indication of the flux density that is reached in the various parts of the structure, highlighting the iron saturation.

Similarly, flux density map gives a prompt view of the field solution, allowing to detect gross and striking mistakes in the field problem setting.

4.2 Point quantities

In each point of the structure, it is possible to get the value of various electric and magnetic quantities. Fig. 7(b) shows some quantities that can be detected in a generic point P of the structure: magnetic vector potential A_z, magnitude and components of flux density vector,

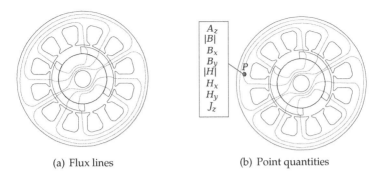

(a) Flux lines   (b) Point quantities

Fig. 7. Flux lines and and magnetic quantities detected in the point P of the solved structure.

$|B|$, B_x and B_y, magnitude and components of field strength vector, $|H|$, H_x and H_y, current density J_z.

4.3 Magnetic flux

The magnetic flux crossing the air gap is computed by integrating along a line in the middle of the air gap the radial flux density (i.e., the normal component to the line) and multiplying by the z-axis length, that is,

$$\Phi = L_{stk} \int_{line} \mathbf{B} \cdot \mathbf{n} dl \tag{4}$$

The integration line is shown in Fig. 8(a), starting from point S (start) and ending to point E (end).

As an alternative to the line integration, the magnetic flux can be achieved from the magnetic vector potential [2]. In this case, it is necessary to compute A_z only in the two ends of the line, i.e., in points S and E. The flux results as

$$\Phi = L_{stk}(A_{z,S} - A_{z,E}) \tag{5}$$

4.4 Flux linkages

The flux linkage of the phase a is computed referring to the coils arrangement reported in Fig. 6(a). The positive coil sides of the phase a are in the slots 3 and 4, while the negative coil sides are within the slots 9 and 10, as highlighted in Fig. 8(b).

---

[2] It is a direct consequence of the Stoke's theorem. Considering the line l_S bordering the surface S, since $\mathbf{B} = curl\mathbf{A}$, it is

$$\int_S \mathbf{B} \cdot \mathbf{n} \, dS = \oint_{l_S} \mathbf{A} \cdot \mathbf{t} \, dl$$

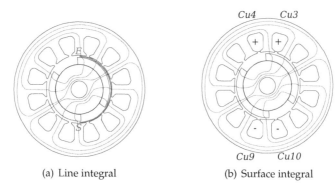

(a) Line integral   (b) Surface integral

Fig. 8. Magnetic flux computation from a line integration, and by means of surface integral in the slots containing the coil sides of phase a.

Using the slot matrix defined in section 3.4, the flux linkage with the phase a results in

$$\Lambda_a = n_{cs} L_{stk} \frac{1}{S_{slot}} \sum_{q=1}^{Q} k_{a,q} \int_{S_{slot,q}} A_z dS \tag{6}$$

where $k_{a,q}$ is the element of the matrix that describes the distribution of the coil sides of the phase a within the q–th slot (see Section 3.4), S_{slot} is the cross–area of the slot.

4.5 Magnetic energy

There are three energy quantities that can be computed over the whole structure. The integral of the product of current density by magnetic vector potential

$$W_{AJ} = L_{stk} \int_{S_{all}} A_z J_z dS \tag{7}$$

Of course, the last integral can be limited in the conductive parts S_c of the structure where the current density is not zero. The magnetic energy is

$$W_m = L_{stk} \int_{S_{all}} \left(\int H dB \right) dS \tag{8}$$

The magnetic coenergy is

$$W_{mc} = L_{stk} \int_{S_{all}} \left(\int B dH \right) dS \tag{9}$$

Even if the magnetic coenergy has not an immediate physical meaning, it is very useful in the magnetic analysis of the machine. Particular care is to define a proper magnetic coenergy density in the PMs which can be defined as

$$w_{mc} = \int_0^{H_m} B_m \, dH_m \tag{10}$$

that results in a negative value. Alternatively, the magnetic coenergy density in the PMs can be also defined as

$$w_{mc} = \int_{H_c^*}^{H_m} B_m \, dH_m \tag{11}$$

that is a positive coenergy. Since the lower limit of integration does not affect the computation of the exchange of magnetic coenergy, the "ideal" magnet characteristic is considered and $H_c^* = B_{rem}/(\mu_{rec}\mu_0)$ is used as the lower limit. Anyway, the rate of change of the magnetic coenergy is the same regardless to the two definitions above, because the difference between the two coenergy densities is the constant quantity $\frac{1}{2}H_c^* B_{rem}$.

If the linear recoil line (1) is adopted for the PM, the magnetic coenergy density results in

$$w_{mc} = \frac{1}{2}\mu_{rec}\mu_0 (H_m - H_c^*)^2$$
$$= \frac{1}{2\mu_{rec}\mu_0} B_m^2 \tag{12}$$

The second form is particularly advantageous since it could be extended to any material type of the system: hard and soft magnetic material as well as non magnetic material.

5. Magnetic analysis example

The magnetic finite element (FE) analysis is applied to a synchronous PM machine with $Q=24$ slots and $p=2$ pole pairs (Slemon & Straughen, 1980) . Thanks to the machine symmetry, only one pole of the machine is analyzed.

5.0.1 Alignment of the rotor with the stator

The rotor angle is fixed to $\vartheta_m^e = 0$ when the d–axis (i.e., the rotor reference axis) is coincident with the a–phase axis (i.e., the stator reference axis). This is represented in Fig. 4(b).

Before starting the analysis under load, it is mandatory to adopt the correct reference, that is, to know the correct position of the rotor with respect the stator reference frame.

From the space vector of the PM flux linkage (that is, at no load) the vector angle α_λ^e is computed, as shown in Fig. 9. If α_λ^e=0 the rotor is in phase with the stator, otherwise the rotor has to be rotated of a mechanical angle corresponding to the electrical angle of the flux linkage vector, that is:

$$\vartheta_m = -\frac{\alpha_\lambda^e}{p} \tag{13}$$

5.1 No load operation

Fig. 10(a) shows one pole of a 24–slot four–pole IPM machine. Fig. 10(b) shows the flux lines at no load, that is, due to the PM only. The flux lines remain in the iron paths of rotor and stator. Some flux lines cross the iron bridges so as to saturate them.

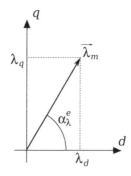

Fig. 9. Flux linkage space vector and its components.

The flux due to the PM is linked by stator winding. It varies according to the position of the rotor. The flux linkage of the phase a is computed, as in (6). Similarly the flux linkages of the phases b and c are computed.

(a) geometry   (b) no–load

Fig. 10. Flux lines in an IPM machine at no load.

Then, the d– and q–axis flux linkages are computed adopting the Park transformation. For convention, at no–load there is only d–axis flux linkage, while the q–axis flux linkage is equal to zero. Therefore, at no load, the PM flux linkage is $\Lambda_m = \Lambda_d$.

The instant in which the d–axis is aligned to the a–axis represents a particular case. The phase a links the maximum flux due to the PM, and thus $\Lambda_m = \Lambda_a$. Therefore the PM flux linkage corresponds to the maximum flux linkage of each phase.

The computation can be repeated varying the rotor position.

5.1.1 Computation of the voltage under load

From the flux linkage waveform, it is possible to compute the induced voltage (also called back EMF) waveform, assuming a constant mechanical speed ω_m (i.e. electrical speed $\omega = p\omega_m$). According with the machine convection, it is

$$e(t) = \frac{d\lambda(t)}{dt} = \frac{d\lambda(\vartheta_m^e)}{d\vartheta_m^e}\omega \tag{14}$$

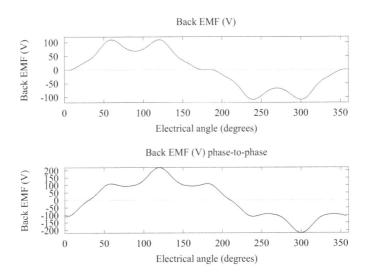

Fig. 11. No load phase and phase-to-phase EMF versus rotor position (at fixed speed).

The flux linkage waveform is expressed by means of its Fourier series expansion. Then, each harmonic of the series is derived, and all harmonics of EMF are summed together. The final EMF waveform is achieved. An example is reported in Fig. 11. A high distortion in the back EMF waveform is evident in this case.

5.1.2 Flux density in the iron

From the field solution, it is important to verify the maximum flux density in the stator teeth and the maximum flux density in the stator back iron. There are two alternative methods.

The first consists in drawing two circle lines centered in the origin of the axis, one crossing the teeth and the other in the middle of the back iron. The maximum flux density is checked along these two circles.

The second method is based on the hypothesis that the flux density in the stator teeth has essentially radial direction, and the flux density in the stator back iron has essentially azimuthal direction. Therefore, the average flux density in each teeth or in each portion of the back iron can be achieved from the vector magnetic potential in the slots.

As an example, the average flux density in the tooth between slot 1 and 2 is

$$B_t = \frac{(A_{z,slot1} - A_{z,slot2})}{w_t} \tag{15}$$

The average flux density in the back iron portion above the slot 1 is

$$B_{bi} = \frac{(A_{z,slot1})}{h_{bi}} \tag{16}$$

$2p$	2 2 2 2 4 4 4 8 8 8
Q	3 6 9 12 6 9 12 6 9 15
GCD	1 2 1 2 2 1 4 2 1 1
N_p	2 1 2 1 2 4 1 4 8 8

Table 2. Number N_p of cogging torque periods per slot pitch rotation

since $A_z = 0$ along the external circumference of the stator, where w_t is the tooth width and h_{bi} is the back iron height. In order to evaluate the maximum flux density, the computations above have to be repeated for each slots.

5.1.3 Cogging torque

Fig. 12 shows the no–load torque versus the rotor position. This torque is due to the interaction between the rotor PM flux and the stator anisotropy due to the teeth, and it is commonly called cogging torque.

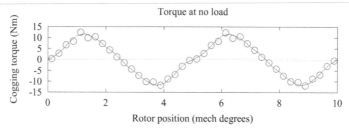

Fig. 12. Cogging torque versus rotor position.

In a rotor with identical PM poles equally spaced the number of N_p periods during a slot pitch rotation is given by

$$N_p = \frac{2p}{GCD\{Q, 2p\}} \tag{17}$$

where GCD means Greater Common Divisor. Thus, the mechanical angle corresponding to each period is $\alpha_{\tau_c} = 2\pi/(N_p Q)$. Higher number of N_p periods lower the amplitude of the cogging torque (Bianchi & Bolognani, 2002). Table 2 reports the values of N_p for some common combinations of Q and $2p$.

5.1.4 Skewing

Skewing rotor PMs, or alternatively stator slots, is a classical method to reduce the cogging torque. In PM machines, the skewing is approximated by placing the PM axially skewed by N_s discrete steps (stepped skewing), as illustrated in Fig. 13. When the stepped skewing is adopted, the FE analysis has to be repeated for each of the N_s section of the machine, considering the correct angle between the stator and the rotor.

Adopting a stepped skewing, there is also a reduction of the back EMF harmonics and the torque ripple.

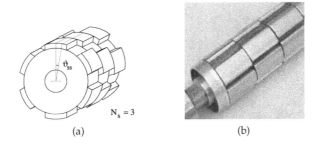

(a)   (b)

Fig. 13. Stepped rotor skewing with three modules

5.2 Operation with stator currents

The analysis is carried out referring to d–q axis component of each electrical and magnetic quantity.

5.2.1 Operation with q–axis stator current

The q–axis current produces a flux in quadrature to the flux due to the PM. When the IPM machine is supplied by q–axis current only, the flux lines go through the rotor and the flux–barrier does not obstruct the q–axis flux, as shown in Fig. 14(a) for a rotor without PM. The q–axis inductance L_q exhibits a high value.

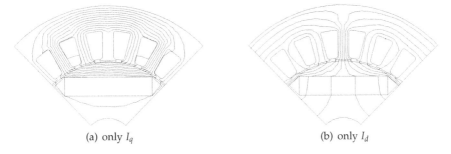

(a) only I_q (b) only I_d

Fig. 14. Flux lines in an IPM machine with (a) I_q only and (b) I_d only.

5.2.2 Operation with d–axis stator current

The d–axis current produces a flux along the d–axis, and thus the flux lines cross the flux–barrier. A positive d–axis current is magnetizing, increasing the flux produced by the PM. Conversely, a negative d–axis current is demagnetizing, since it weakens the PM flux.

Fig. 14(b) shows the flux plot due to the d–axis current I_d only (is case of a rotor without PMs). The flux lines are similar to the flux lines due to PMs, shown in Fig. 10(b). The d–axis inductance L_d results lower than L_q. The analytical estimation of L_d is tightly dependent on the geometry of the flux–barriers (Bianchi & Jahns, 2004) . The ratio between the two inductances, i.e. $\xi = L_q / L_d$ is the saliency ratio (Miller, 1989) .

5.2.3 Magnetic model of the PM synchronous machine

The magnetic model is commonly expressed in the synchronous d–q reference frame. The relationship between the d– and the q–axis flux linkages and currents is non linear, as shown in Fig. 15. It is given by

$$\lambda_d = \lambda_d(i_d, i_q)$$
$$\lambda_q = \lambda_q(i_d, i_q)$$
(18)

They are single–valued functions, because it is assumed that energy stored in the electromagnetic fields can be described by state functions (White & Woodson, 1959). Such a model is used for an accurate estimation of the machine performance: to precisely predict the average torque, the torque ripple, the capability to sensorless detect the rotor position.

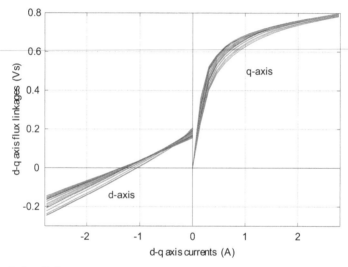

Fig. 15. d–q axis flux linkages versus currents.

From the d– and q–axis flux linkages, the d– and q–axis voltages are

$$v_d = Ri_d + \frac{d\lambda_d}{dt} - \omega\lambda_q$$

$$v_q = Ri_q + \frac{d\lambda_q}{dt} + \omega\lambda_d$$
(19)

Assuming a linear characteristic of all iron parts, the d– and q–axis flux linkages simplify as

$$\lambda_d = \Lambda_m + L_d i_d$$

$$\lambda_q = L_q i_q$$
(20)

and the voltage components result in

$$v_d = Ri_d + L_d\frac{di_d}{dt} - \omega L_q i_q$$

$$v_q = Ri_q + L_q\frac{di_q}{dt} + \omega(\Lambda_m + L_d i_d)$$

(21)

Fig. 16 shows the steady–state [3] vector diagram of the PM synchronous machine in d–q reference frame (Boldea & Nasar, 1999).

6. The electromechanical torque

Adopting the FE model of the machine, the torque T can be computed by integrating the Maxwell stress tensor along the rotor periphery:

$$T = \frac{D^2 L_{stk}}{4} \int_0^{2\pi} \frac{B_{g,n} B_{g,\theta}}{\mu_0} d\theta_m$$

(22)

where $B_{g,n}$ and $B_{g,\theta}$ are the normal and tangential component of the air gap flux density, and ϑ_m is the rotor position (Ida & Bastos, 1992; Jin, 1992; Salon, 1995). However, assuming the rotor position ϑ_m and d– and q–axis currents i_d and i_q as state variables, the machine torque is also given by

$$T = \frac{3}{2}p\left(\lambda_d i_q - \lambda_q i_d\right) + \frac{\partial W_{mc}}{\partial \vartheta_m}$$

(23)

where p is the number of pole pairs. W_{mc} is the magnetic coenergy, which is a state function of ϑ_m, i_d and i_q, i.e. $W_{mc} = W_{mc}(\vartheta_m, i_d, i_q)$ (White & Woodson, 1959). The first term of the second member of (23) is labeled as T_{dq}, that is

$$T_{dq} = \frac{3}{2}p(\lambda_d i_q - \lambda_q i_d).$$

(24)

This torque term T_{dq} is slightly affected by the harmonics of the flux linkages and it results to be suitable for the computation of the average torque. The torque ripple is mainly described by the coenergy variation, that is the second term of (23).

6.0.4 Cogging torque

Cogging torque is the ripple torque due to the interaction between the PM flux and the stator teeth. Since the stator currents are zero, it is T_{dq}=0 and, from (23), the cogging torque results to be equal to

$$T_{cog} = \frac{\partial W_{mc}}{\partial \vartheta_m}$$

(25)

Fig. 12 shows the cogging torque versus rotor position of an SPM machine. Solid line refers to the torque computation by means of the Maxwell stress tensor while the circles refer to the

---

[3] Time derivatives are equal to zero and thus the currents are constant, $i_d = I_d$ and $i_q = I_q$.

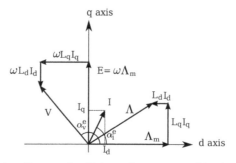

Fig. 16. Steady–state vector diagram for PM synchronous machine in $d - q$ reference frame

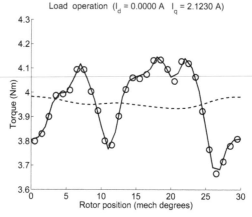

Fig. 17. Torque behavior under load of the SPM machine. Solid line refers to the computation using Maxwell stress tensor, dashed line refers to (24), circles refer to (23).

torque computation (25). A further comparison between predictions and measurements of cogging torque is reported in (Bianchi & Bolognani, 2002) .

6.0.5 Computation under load

Fig. 17 shows the torque behavior versus rotor position of the SPM machine fed by q–axis current only, while d–axis current is zero. Solid line refers to the Maxwell stress tensor computation. The circles refer to the torque computation (23). The dashed line refers to the torque computation T_{dq}, given by (24). As said above, the behaviour of T_{dq} is smooth and close to the average torque. Similar results are found when an IPM machine is considered, as the IPM machine shown in Fig. 1(c). Some torque behaviors and experimental results are reported in (Barcaro et al., 2010A;C) .

6.1 Optimizing the torque behaviour

One of the purposes of the FE analysis is to optimize the torque of the machine: high average torque and limited torque ripple (Fratta et al., 1993) . Many strategies exist so as to reduce the torque ripple, and FE analysis is a proper tool to optimize the machine geometry (Bianchi & Bolognani, 2000; Sanada et al., 2004) .

Unusual geometries can be analyzed in order to improve the machine performance. As an example, Fig. 18(a) shows an SPM rotor in which the PMs have been shifted with respect their symmetrical position, so as to reduce the PM flux linkage and the torque harmonics (Bianchi & Bolognani, 2000; 2002) . Fig. 18(b) shows the "Machaon" [4] rotor, proposed to reduce the torque ripple in IPM machines (Bianchi et al., 2008; 2009) . It is formed by laminations with flux–barriers of different geometry, large and small alternatively under the adjacent poles.

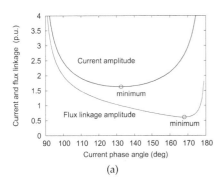
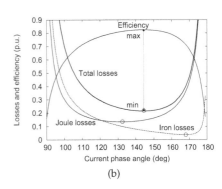

(a) PM shifting   (b) Machaon rotor

Fig. 18. Photos of proper solutions to optimize the torque behavior.

6.2 Searching the MTPA trajectory

A proper control strategy is mandatory to achieve high performance of the PM synchronous motors. Fig. 19 shows the key characteristics of the machine as a function of the current vector angle α_i^e, defined in Fig. 16, for given torque and speed. Normalized parameters are used, that is, unity torque $\tau_{pu} = 1$ and unity speed $\omega_{pu} = 1$ are fixed.

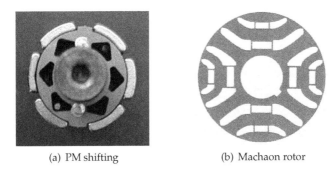

Fig. 19. Stator current, flux–linkage, losses and efficiency as a function of current vector angle α_i^e under constant torque ($\tau_{pu} = 1$) and constant speed ($\omega_{pu} = 1$) condition.

---

[4] The name comes from a butterfly with two large and two small wings.

For a given torque, there is an optimal operating point in which the current is minimum, as shown in Fig. 19(a). Therefore, a maximum torque to current ratio exists. When such a ratio is maximized with respect the current vector angle α_i^e for any operating condition, the maximum torque–per–Ampere (MTPA) control is achieved (Jahns et al., 1986) .

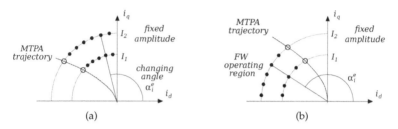

(a)   (b)

Fig. 20. Procedure of changing the current vector angle α_i^e, searching the MTPA trajectory and defining the flux–weakening capability.

The FE procedure to search the maximum torque per Ampere is illustrated in Fig. 20(a):

- the current amplitude is fixed,
- the current vector angle α_i^e varies from 90 to 180 electrical degrees[5],
- the torque is computed for each current vector.

The flux linkage corresponding to the point of maximum torque is the base flux linkage Λ_B.

6.3 The flux–weakening operating region

Similarly, the flux–weakening capability of the machine can be investigated starting from the MTPA trajectory, as shown in Fig. 20(b). For each current vector the torque and the d– and q–axis flux linkage components are computed. According to the flux linkage amplitude, Λ, imposing the voltage limit, V_n, the maximum electrical speed is computed as $\omega = V_n/\Lambda$. Then, the corresponding mechanical speed $\omega_m = \omega/p$ is related to the electromagnetic torque.

Repeating the computation for different current vector angle α_i^e, the whole characteristic torque versus speed is achieved. Fig. 21(a) shows the current trajectory, and Fig. 21(b) shows the corresponding torque versus speed curve and power versus speed curve.

7. Prediction of sensorless capability of PM motors

The technique based on the high–frequency voltage signal injection is used for sensorless rotor position detection of PM synchronous machines at zero and low speed (Ogaswara & Akagi, 1998) . It is strictly bound to the rotor geometry, requiring a synchronous PM machine with anisotropic rotor, e.g. an IPM machine as in Fig. 1(c) or an inset machine as in Fig. 1(b).

A high–frequency stator (pulsating or rotating) voltage is added to the fundamental voltage, then the corresponding high–frequency stator current is affected by the rotor saliency (Harke et al., 2003; Linke et al., 2003) and information of the rotor position is extracted from current measurement (Consoli et al., 2000; Jang et al., 2003) .

---

[5] Actually, the search ends when the torque starts to decrease.

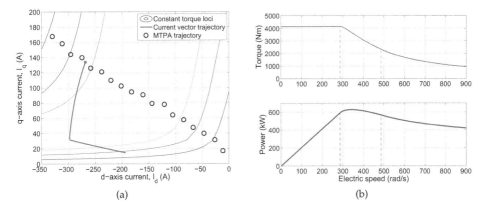

Fig. 21. Current trajectory (a) and torque and power versus speed (b).

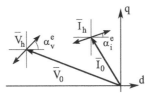

Fig. 22. Phasor diagram with steady–state and high frequency components

An accurate magnetic model of the machine is mandatory to predict the capability of the machine for the sensorless rotor position detection considering both saturation and cross–saturation effect.

The magnetic model to predict the error signal $\varepsilon(\vartheta^e_{err})$ is achieved by a set of finite element simulations carried out so as to compute the d– and q–axis flux linkages as functions of the d– and q–axis currents (Bianchi, 2005) .

Then, for a given operating point (defined by the fundamental d– and q–axis currents), a small–signal model is built, defined by the incremental inductances

$$L_{dd} = \frac{\partial \lambda_d}{\partial i_d} \qquad L_{dq} = \frac{\partial \lambda_d}{\partial i_q} \qquad L_{qd} = \frac{\partial \lambda_q}{\partial i_d} \qquad L_{qq} = \frac{\partial \lambda_q}{\partial i_q}$$

When a high frequency voltage vector is injected along the direction α^e_v (i.e. the d– and q–axis voltage components are $V_h \cos \alpha^e_v$ and $V_h \sin \alpha^e_v$ respectively), the small–signal model (26) allows to compute the amplitude and the angle of current vector.

The phasor diagram is shown in Fig. 22, including both fundamental components ($\overline{V}_0$ and $\overline{I}_0$) and high frequency components ($\overline{V}_h$ and $\overline{I}_h$).

Such a study is repeated, varying the voltage vector angle α^e_v, so as to estimate the rotor position error signal ε. Let I_{max} and I_{min} the maximum and minimum of the high frequency current (computed with the various voltage vector angles α^e_v), and α^e_{Imax} the angle where I_{max} is found (defined with respect to the d–axis). Then, the rotor position estimation error signal is computed as

$$\varepsilon(\alpha_v^e) = k_{st} \frac{I_{max} - I_{min}}{2} \sin 2(\alpha_v^e - \alpha_{Imax}^e) \tag{26}$$

where α_v^e can be considered as the injection angle and $(\alpha_v^e - \alpha_{Imax}^e)$ can be considered as the error signal angle ϑ_{err}^e. Then, α_{Imax}^e corresponds to the angular displacement due to the d–q axis cross–saturation.

Fig. 23 compares experimental and predicted results, referring to the inset machine shown in Fig. 1(b), whose rated current is $\hat{I} = 2.5$ A. The pulsating voltage vector technique has been used, adopting a high–frequency voltage with amplitude $V_h = 50$ V, frequency $f_c = 500$ Hz, and a machine speed $n = 0$ rpm (Bianchi et al., 2007). The satisfactory match between predictions and measurements, confirms that a PM machine model can be profitably used to predict the sensorless capability of the machine.

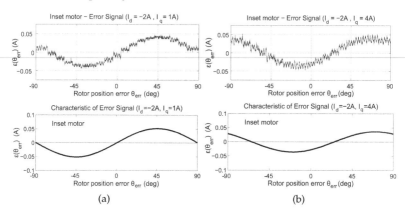

Fig. 23. Rotor position error ϑ_{err}^e and estimation error signal ε with pulsating voltage injection and inset PM machine (experimental test and prediction).

8. Losses in electrical machines

The computation of the losses in an electrical machine is a complex task, sometimes requiring an involved model of the machine even though some uncertainties (e.g. about the material used) prevent a precise estimation. Therefore, in many cases, simpler loss estimations are quite adequate to predict the machine capabilities.

8.1 Stator winding losses

Stator resistance is a three–dimensional parameter of the electrical machine. It is generally computed analytically on the basis of the wire diameter and the total length of the winding, including stack length and end–winding length.

The copper conductivity is decreased according to the temperature. Such a temperature is obtained and adjusted after the thermal computation. Sometime an analysis loop is necessary.

Joule losses are computed multiplying the resistance by the square of the rms value of the stator current: $P_{js} = 3R_s I_s^2$. When the current includes time harmonics, the equivalent root–mean–square (rms) current is calculated as:

$$I_{rms} = \sqrt{I_0^2 + \frac{\hat{I}_1^2 + \hat{I}_2^2 + \cdots + \hat{I}_n^2}{2}} \tag{27}$$

where I_0 is the constant value (if any), and $\hat{I}_1 \ldots \hat{I}_n$ are the Fourier series coefficients (peak value) of the current waveform.

8.2 Stator iron losses

The iron losses of a generic motor consist of the sum of the hysteresis loss, classical eddy current loss and excess loss. Considering a magnetic flux density $\hat{B}$ varying sinusoidally at the frequency f, the iron loss density is commonly expressed in the following form (Boglietti et al., 2003) as

$$p_{iron} = k_{hy}\, \hat{B}^\beta\, f + k_{ec}\, \hat{B}^2\, f^2 + k_{ex}\, \hat{B}^{\frac{3}{2}}\, f^{\frac{3}{2}} \tag{28}$$

where k_{hy} and k_{ec} are the hysteresis and the classical eddy current constant, and β is the Steinmetz constant, often approximated as $\beta \simeq 2$. The k_{ex} is the eddy current excess losses constant. These losses are due to the dynamic losses of the Weiss domains when a variable magnetic field is applied to the magnetic material. The block walls discontinuous movements produce fast Barkhausen jumps and then eddy currents.

These constants should be obtained from material data sheet, but the Epstein frame test does not allow to segregate between the eddy currents due to the classical losses from the eddy currents due to the excess losses. Even if the excess current loss component could be very significant in many lamination materials (Bertotti, 1998), in (Boglietti et al., 2003) the difficulty to separate the excess losses contribution from the classical eddy current losses has been highlighted. A single eddy current losses coefficient is defined, i.e. an increased k_{ec}, and then applied neglecting the third addendum of (28).

Since the stator iron teeth and the stator back iron operate at different flux density values, the iron loss density has to be computed separately referring to the two parts of the stator.

8.3 Tooth iron losses computation with distorted flux density

The equations above hold for sinusoidal flux density variations. When the flux density varies in the iron paths with different waveforms, the computation of the stator iron losses is more complex.

The flux distortions are mainly located in the stator teeth. Let B_t the tooth flux density computed as a function of the time (or of the rotor position, when the speed is considered to be constant). An example of the B_t waveform is shown in Fig. 24, referring to a four–pole 24–slot machine. Since the machine has two slots per pole per phase, therefore two teeth are considered. It is worth noticing that the waveforms are quite different from sinusoidal waveforms. The behavior of the tooth flux density is expressed by means of Fourier series expansion:

$$B_t(\theta) = \sum \hat{B}_h \sin(h\theta + \alpha_h) \tag{29}$$

where h is the order of the harmonic of the tooth flux density.

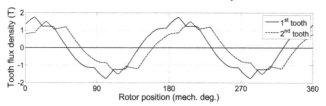

Fig. 24. Tooth flux density versus rotor position.

Only eddy current losses are considered since they are the greatest part of total iron losses due to the flux density harmonics, being proportional to f^2, see (28).

These losses are due to fluctuation of the tooth flux density. Then the tooth eddy current iron loss density is given by:

$$p_{ec} = \frac{k_{ec}}{2\,\pi^2} \frac{\omega^2}{2\,\pi} \int_0^{2\pi} \left(\frac{\partial B_t}{\partial \theta}\right)^2 d\theta \qquad (30)$$

And after some manipulations, it is

$$p_{ec} = k_{ec}\, f^2 \sum \hat{B}_h^2\, h^2 \qquad (31)$$

In (Barcaro et al., 2010B) the effect of the IPM rotor structure on the tooth eddy current iron loss density has been analyzed.

8.4 Rotor losses due to MMF harmonics

The discrete location of the coils within the stator slots causes space harmonics of the magneto motive force (MMF) traveling in the air gap. These MMF harmonics move asynchronously with respect to the rotor inducing currents in any conductive rotor parts (Atallah et al., 2000; Shah & Lee, 2006). The losses in the rotor volume due to the induced currents are given by:

$$P_{rl} = \int_{vol} \frac{J_r^2}{\sigma} dVol = \int_{vol} \sigma \frac{\partial A_z}{\partial t} \qquad (32)$$

where J_r is the current density induced in the rotor, A_z is the magnetic vector potential, σ is the material conductivity and t is the time.

These losses increase rapidly with the machine size. When the scaling law is apply the flux density B results to be proportionally to the linear quantity l, the curl of B and the induced current density J_r to l^2. Since the volume increases as l^3, the rotor losses result to be proportionally to l^7. Although these equations above do not consider skin effects or iron saturation, they highlight how the rotor losses might increase with the size of the machine. The rotor losses phenomenon can be neglected in small size PM machine, it has to be considered compulsorily in large size PM machines.

A practical finite element computation of rotor losses is based on the superposition of the effects: the total rotor losses P_{rl} result as the sum of the rotor losses computed for each harmonic order ν, that is $P_{rl} = \sum_\nu P_{rl,\nu}$. To this aim, linear iron is considered assuming an equivalent permeability or freezing the magnetic permeability after a magnetostatic field solution (Bianchi & Bolognani, 1998). For each v-th MMF harmonic, the following procedure is adopted:

(1) The stator is substituted by an infinitesimal sheet placed at the stator inner diameter D, as shown in Fig. 25. A linear current density $K_{sv}(\vartheta)$, sinusoidally distributed in space, is imposed in such a stator sheet:

$$K_{s,v}(\theta) = \hat{K}_{s,v} \cdot \sin(v\theta + \omega_{vr}t) \tag{33}$$

where v is the harmonic order, ω_{vr}/v is the speed of such harmonic with respect to the rotor, and $\hat{K}_{sv}$ is the peak value of linear current density, achieved from the corresponding MMF harmonic $\hat{U}_{sv}$. They are:

$$\omega_{vr} = \left(\frac{\omega}{sgn \cdot v} - \frac{\omega}{p} \right) \tag{34}$$

$$\hat{K}_{sv} = 2v\frac{\hat{U}_{sv}}{D} \tag{35}$$

where sgn is equal to $+1$ or -1 according to whether the harmonic speed is forward or backward the rotor speed (Bianchi & Fornasiero, 2009).

(2) The circumference is split in a high number of points, e.g. N_p. In each point a prefixed *point current* I_{pv} is assigned, as shown in Fig. 25(b), which is the integral of the distribution of the linear current density over an arc length $\pi D/N_p$. According to the v-th harmonic, the maximum current value is computed from the electric loading $\hat{K}_{sv}$, as $\hat{I}_{pv} = \hat{K}_{sv}\pi D/N_p$. The linear current density waveform rotates along the air-gap at the speed ω_{vr} in the rotor reference frame. The points currents are alternating. Using the symbolic notation, in the generic angular position ϑ, the *point current* is

$$\dot{I}_{pv}(\vartheta) = \hat{I}_{pv}e^{jv\vartheta} \tag{36}$$

where the phase of the current (i.e. $v\vartheta$) is a function of the geometrical position ϑ of the point.

(3) The frequency of the simulation is computed as

$$f_{vr} = f\left(sgn \cdot \frac{v}{p} - 1 \right) \tag{37}$$

In each simulation step, the rotor losses due to a single MMF harmonic are computed (Shah & Lee, 2009). The simulation needs a particular care adopting a two–dimensional analysis. In each object, electrically insulated by the others, a total current equal to zero is imposed as a further constraint. In addition, when laminations are insulated, a conductivity equal to zero is fixed. The iron conductivity is $\sigma_{Fe} = 3MS/m$, and the magnet conductivity is $\sigma_{PM} = 1.16MS/m$. Mesh size is chosen according to the penetration thickness.

According to the 12–slot 10–pole PM machine, with a single–layer winding, Fig. 26(a) shows the amplitudes of the MMF harmonics, as percentage of the main harmonic highlighting the presence of a subharmonic of order $v = 1$. Fig. 26(b) shows their frequency. Fig. 27 shows the flux lines in the rotor of the 12–slot 10–poles PM machine due to the MMF subharmonic ($v = 1$) and other two MMF harmonics of higher order ($v = 7$ and $v = 11$).

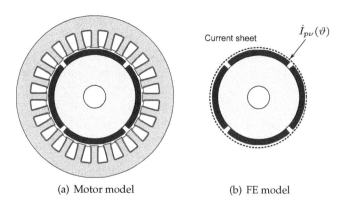

(a) Motor model   (b) FE model

Fig. 25. Model used for FE computations.

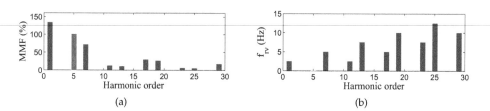

(a)   (b)

Fig. 26. Stator current MMF harmonic amplitudes and corresponding rotor frequency versus harmonic order, according to a 12–slot 10–poles single–layer PM machine.

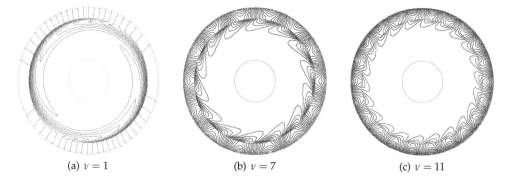

(a) $\nu = 1$ (b) $\nu = 7$ (c) $\nu = 11$

Fig. 27. Flux lines due to MMF harmonics of order $\nu = 1$ (subharmonic), $\nu = 7$ and $\nu = 11$.

9. Temperature rises computation

Once the losses of the electrical machine are estimated, it is possible to compute the temperature rise in various parts of the electrical machine. It requires the knowledge of the thermal properties of the material used in the machine, as well as the heat dissipation conditions with the external environment. A two–dimension analysis is considered hereafter.

9.1 Model of the machine

The FE model of a PM machine is shown in Fig. 28. Each material is characterized by a proper thermal conductivity, and a volume heat generation proportional to its losses. Due to the symmetry of the machine, only a portion of the machine can be analyzed.

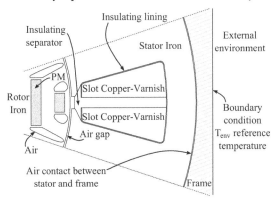

Fig. 28. Detail of machine geometry and material properties adopted in the FE analysis.

When the higher losses are located in the stator, a portion formed by one slot and two half–teeth is enough in the analysis. Fig. 28 shows one slot pitch of an interior PM machine. On the stator external surface, there is an Aluminum frame. Between the external surface of the stator lamination and the Aluminum frame, an air foil is added, so as to take into account the roughness and the imperfect contact between the two surfaces. The thickness of such a foil is in the range 0.02–0.05 mm.

Within the slot, between the stator lamination and the coil, a slot insulating lining is considered. An insulating separator is added in the middle of these two coil layers to increase the electrical insulation between different phases.

9.2 Material thermal properties

The thermal conductivity of the materials are reported in Table 3. As far as the thermal conductivity of the air–gap is concerned, it refers to a fluido–dynamic calculations, using the rotation speed of the machine, as will be described hereafter.

The quality of some materials is dependent on the temperature. Among the others, particular attention is devoted to PMs. Both the residual flux density and the magnetic field of the knee of the $B - H$ curve are reduced, as reported in Fig. 3(b). As said previously the reduction of the magnetic field increases the risk of an irreversible demagnetization of the PMs.

9.2.1 Air gap

The air flow in the air gap is turbulent. The heat transfer is described with an effective thermal conductivity λ_{gap} that is defined as the thermal conductivity that the stationary air should have in order to transfer the same amount of heat as the moving air (Mademlis et al., 2000) .

Material	Symbol	Value ($W/m \cdot K$)
Insulating lining	λ_{ins}	0.15
Insulating separator	λ_{sep}	0.25
Copper+Varnish	λ_{wnd}	0.70—1.0
Aluminum	λ_{Al}	100
Iron	λ_{Fe}	50
Air	λ_{air}	0.026
Air gap	λ_{gap}	0.130
Magnet	λ_{mag}	9
Shaft	λ_{sh}	50

Table 3. Thermal conductivity of the main materials

Its value depends on the relative speed ω_m between stator and rotor as well as the air gap width. It is calculated as

$$\lambda_{gap} = 0.0019 \, \eta^{-2.9084} \, Re^{0.4614 \, \ln(3.3361 \, \eta)} \tag{38}$$

where

$$\eta = \frac{D_i - 2 \, g}{D_i}$$
$$Re = \frac{\omega_m \, g}{\nu} \tag{39}$$

and ν is the cinematic viscosity of air. Obliviously, if the rotor is at standstill, the air gap conductivity is equal to the stationary value, i.e. λ_{air}.

9.2.2 Slot

The actual slot contains many conductors, insulated each other by means of varnish. However, the drawing of all conductors make no sense. Therefore, the coil winding is modeled as an equivalent homogeneous material characterized by a proper thermal conductivity, as shown in Fig. 29.

Fig. 29. Thermal equivalent model of the slot

The equivalent thermal conductivity is computed as suggested in (Mellor et al., 1991; Schuisky, 1967) . It depends on the ratio between the copper diameter d_c and the varnish–insulated diameter d_{ins}. It is

$$\lambda_{Cu-ins} = F \, \lambda_{ins} \tag{40}$$

where λ_{ins} is the varnish–insulating conductivity, and the F is a multiplier factor (Schuisky, 1967) calculated as $F = 37.5\,x^2 - 43.75\,x + 14$ $(x = d_c/d_{ins})$. Typical value of such a conductivity is 4 to 5 times that of the insulation, for the wire dimensions adopted in practice.

9.3 Assigning the loss sources

In each part of the electrical machine in which the losses are generated, a heat generation is imposed as a heat source. Therefore, each region of the model is characterized by specific losses per volume (in W/m^3).

The specific losses for the region within the stator slots are computed dividing the Joule losses of the stator winding by the volume of the stator slots. The length L_{stk} of the model has to be considered.

The specific losses for the stator iron region correspond to the stator iron losses divided by the stator iron volume. When the iron tooth losses are quite different from the back iron losses, it is convenient to consider two different regions where to assign the specific losses. In the PM and rotor iron regions, the specific rotor losses are assigned.

9.4 Assign the boundary conditions

The temperatures of the winding in the slots, the stator iron and the PMs are referred to the environment temperature, which is fixed to be $T_{env} = 0°C$. Therefore, the field solution yields the temperatures rise with respect to the environment temperature.

A natural air convection is considered externally. This yields a thermal convection coefficient equal to 6 $W/(m^2K)$. However, in the two–dimensional FE model, a higher thermal convection coefficient is set taking into account of the surface increase due to the external length of the frame as respect to the stator stack length, and the presence of fins, which are not detailed in the model of Fig. 28. Typically a factor 3 is used, yielding the thermal convection coefficient to be 18 $W/(m^2K)$. This is the boundary conditions at the outer surface of the machine frame.

9.5 Thermal computation

Fig. 30(b) shows the result of the thermal FE analysis. The steady–state temperature rises are indicated in the slots, in the PMs and in the stator iron. The thermal analysis refers to rated conditions, considering only the Joule losses (standstill operations).

The higher temperature is reached in the slots. The maximum temperature rise reached is about 90 K with respect to the environment temperature. The experimental measurements (Barcaro et al., 2011) confirm the simulated results. According to an external temperature of 20 °C, the winding temperature reaches 110 °C, and the frame temperature 91 °C.

9.6 Overload and faulty operating conditions

Overload operations are typically required in many applications, where the operating mode is discontinuous and repetitive accelerations and decelerations are demanded to the electrical

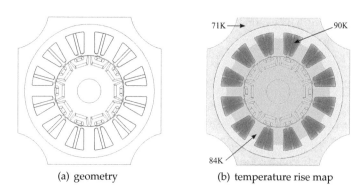

(a) geometry   (b) temperature rise map

Fig. 30. Machine geometry and simulated temperature rise distribution over environment temperature in healthy mode.

machine. Therefore, the electrical machine has to be designed so as to allow temporary overload operations, according to the given cooling system of the system.

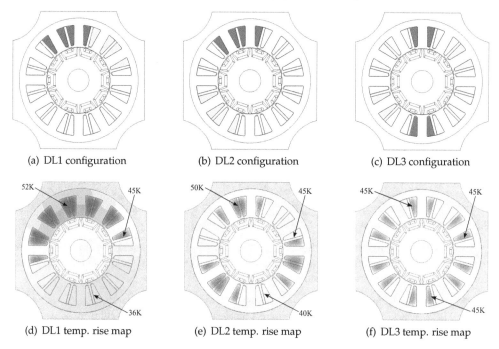

(a) DL1 configuration   (b) DL2 configuration   (c) DL3 configuration

(d) DL1 temp. rise map   (e) DL2 temp. rise map   (f) DL3 temp. rise map

Fig. 31. Machine geometry and simulated temperature rise distribution over environment temperature in faulty mode: (a) and (d) DL1 configuration, (b) and (e) DL2 configuration, (c) and (f) DL3 configuration.

Particular analysis deals with the faulty operating conditions, in which only a portion of the machine is operating, while the other is disconnected from the power supply.

Hereafter, a dual–three–phase machine is considered in which only half a coils are supplied and the others are open circuited. The machine is fed by a phase current higher than the nominal value, still satisfying the thermal insulating class limit. Referring to the 12–slot 10–pole double–layer PM machine this post–fault strategy has been proposed in (Barcaro et al., 2011) .

9.6.1 Torque under faulty condition

The thermal analysis allows to compute which is the maximum torque that a machine can exhibit according to the limit temperature rise and the fixed cooling system. Some results are presented hereafter referring to the dual–three–phase machine, during faulty operating conditions. Three different coil connections are considered. They are labeled as DL1, DL2, and DL3, according to how the two sets of three–phase windings are placed within the stator.

Fig. 31(d) shows the temperature map in case of one open–circuited phase according to the DL1 configuration of Fig. 31(a). Since the temperature rise of the winding is 52 K (in comparison with 90 K of healthy operating conditions) the operating current could be increased by a factor of $\sqrt{90/52} = 1.32$.

Therefore, according to the thermal analysis, when the dual–three–phase machine is operated under faulty operating conditions, it is possible to reach a torque about 70% of the healthy value overloading the machine and without exceeding the limit temperature.

Fig. 31(e) shows the temperature map in case of one open–circuited phase according to the DL2 configuration of Fig. 31(b). Fig. 31(f) shows the temperature map in case of one open–circuited phase according to the DL3 configuration of Fig. 31(c).

9.7 Impact of the rotor losses

The rotor losses can have a strong impact on the temperature rises of the machine. Fig. 32 shows the temperature map for a PM machine and neglecting and considering rotor losses. Colors from light blue to dark red show the temperature rise in the machine. In this example, the temperature rises in PMs and Copper increase from 80 K to 100 K, and from 90 K to 98 K, respectively. It is worth noticing how the rotor losses influence the thermal behavior of the two machines. There is an evident increasing of the temperature both in PMs and in the winding, that can cause magnet demagnetization as shown in Fig. 3(b).

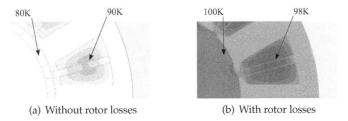

(a) Without rotor losses   (b) With rotor losses

Fig. 32. Effect of the rotor losses in the machine temperature

This means that the rotor losses can not be neglected, in order to avoid an underestimation of the operating temperatures.

10. Acknowledgement

The authors thank many collegues of theirs for their help and their suggestions in collecting the material for this chapter. Among the others, let's remember PhD. Michele Dai Pré, Eng. Diego Bon, PhD. Luigi Alberti, PhD. Emanuele Fornasiero, Eng. Mosé Castiello, Eng. Alessandro Fasolo, and Eng. Dario Durello.

11. References

Atallah, K., Howe, D., Mellor, P. H., & Stone, D. A. (2000). Rotor loss in permanent-magnet brushless AC machines, *IEEE Transactions on Industry Applications*, Vol. 36 (No. 6): 1612–1618.

Barcaro, M., Alberti, L., Faggion, A., Sgarbossa, L., Dai Pre, M., Bianchi, N. & Bolognani, S. (2010). IPM Machine Drive Design and Tests for an Integrated Starter-alternator Application, *IEEE Transactions on Industry Applications*, Vol. 46 (No. 3): 993–1001.

Barcaro, M., Bianchi, N., & Magnussen, F. (2010). Rotor Flux-Barrier Geometry Design to Reduce Stator Iron Losses in Synchronous IPM Motors Under FW Operations, *IEEE Transactions on Industry Applications*, Vol. 46 (No. 5): 1950–1958.

Barcaro, M., Bianchi, N., & Magnussen, F. (2010). Analysis and tests of a dual three-phase 12-slot 10-pole permanent-magnet motor, *IEEE Transactions on Industry Applications*, Vol. 46 (No. 6): 2355–2362.

Barcaro, M., Bianchi, N. & Magnussen, F. (2011). Faulty Operations of a PM Fractional-Slot Machine With a Dual Three-Phase Winding, *IEEE Transaction on Industrial Electronics*, Vol. 58 (No. 9): 3825–3832.

Bertotti, G. (1998). *Hysteresys in Magnetism: For Physicists, Material Scientists, and Engineers*. Academic Press.

Bianchi, N. & Bolognani, S. (1998). Magnetic models of saturated interior permanent magnet motors based on finite element analysis, *Proc. of Industrial Applications Conference Records (IAS)*, IEEE, S.Louis (USA), pp. 27–34 (Vol. 1).

Bianchi, N. & Bolognani, S. (2000). Reducing torque ripple in PM synchronous motors by pole shifting, *Proc. of International Conference on Electrical Machines (ICEM)*, Helsinki (Finland), pp. 1222–1226.

Bianchi, N. & Bolognani, S. (2002). Design techniques for reducing the cogging torque in surface–mounted PM motors, *IEEE Trans. on Industry Applications*, Vol. 38 (No. 5): 1259–1265.

Bianchi, N. & Jahns, T. (ed.) (2004). *Design, Analysis, and Control of Interior PM Synchronous Machines, Tutorial Course notes of IEEE Industry Applications Society Annual Meeting (IAS)*, CLEUP, Padova (Italy).

Bianchi, N. (2005). *Electrical Machine Analysis using Finite Elements*. CRC Press, Taylor & Francis Group, Boca Raton (USA).

Bianchi, N., Bolognani, S., Jang, J.-H., & Sul, S.-K. (2007). Comparison of PM Motor Structures and Sensorless Control Techniques for Zero–Speed Rotor Position Detection, *IEEE Transactions on Power Electronics*, Vol. 22 (No. 6): 2466–2475.

Bianchi, N., Dai Pré, M., Alberti, L., & Fornasiero, E. (2007). *Theory and Design of Fractional-Slot PM Machines, Tutorial Course notes of IEEE Industry Applications Society Annual Meeting (IAS)*, CLEUP, Padova (Italy).

Bianchi, N., Bolognani, S., Bon, D. & Dai Pré, M. (2008) Torque Harmonic Compensation in a Synchronous Reluctance Motor, *IEEE Transactions on Energy Conversion*, Vol. 23 (No. 2): 466–473.

Bianchi, N. & Fornasiero, E. (2009). Impact of MMF Space Harmonic on Rotor Losses in Fractional-slot Permanent-magnet Machines, *IEEE Transactions on Energy Conversion*, Vol. 24 (No. 2): 323–328.

Bianchi, N., Bolognani, S., Bon, D. & Dai Pré, M. (2009). Rotor Flux-Barrier Design for Torque Ripple Reduction in Synchronous Reluctance and PM-Assisted Synchronous Reluctance Motors, *IEEE Trans. on Industry Applications*, Vol. 45 (No. 3): 921–928.

Boglietti, A., Cavagnino, A., Lazzari, M. & Pastorelli, M. (2003). Predicting iron losses in soft magnetic materials with arbitrary voltage supply: an engineering approach, *IEEE Transactions on Magnetics*, Vol. 39 (No. 2): 981–989.

Boldea, I. & Nasar, S. A. (1999). *Electric Drives*, CRC Press, Taylor & Francis Group, Boca Raton (USA).

Bozorth, R. M. (1993). *Ferromagnetism*, IEEE Press, New York (USA).

Coey, J. C. (ed.) (1996). *Rare Heart Iron Permanent Magnet – Monographs on the Physics and Chemistry of Materials*, Oxford Science Publications, Claredon Press, Oxford (UK).

Consoli, A., Scarcella, G., Tutino, G., & Testa, A. (2000). Sensorless field oriented control using common mode currents, *Proc. of Industry Application Society Annual Meeting (IAS)*, IEEE, Rome (Italy), pp. 1866–1873.

Fratta, A., Troglia, G., Vagati, A. & Villata, F. (1993). Evaluation of torque ripple in high performance synchronous reluctance machines, *Proc. of Industry Application Society Annual Meeting (IAS)*, IEEE, Toronto (Canada), pp. 163–170.

Harke, M., Kim, H., & Lorenz, R. (2003). Sensorless control of interior permanent magnet machine drives for zero–phase–lag position estimation, *IEEE Transaction on Industry Applications*, Vol. IA–39 (No. 12): 1661–1667.

Honsinger, V. (1982). The fields and parameters of interior type AC permanent magnet machines, *IEEE Trans. on PAS*, Vol. 101: 867–876.

Ida, N. & Bastos, J. (1992). *Electromagnetics and Calculation of Fields*. Springer-Verlag Inc, New York (USA).

Jahns, T., Kliman, G., & Neumann, T. (1986). Interior PM synchronous motors for adjustable speed drives, *IEEE Trans. on Industry Applications*, Vol. IA–22 (No. 4): 738–747.

Jang, J., Sul, S., & Son, Y. (2003). Current measurement issues in sensorless control algorithm using high frequency signal injection method, *Proc. of Industry Application Soceity Annual Meeting (IAS)*, IEEE, Salt Lake City (USA), pp. 1134–1141.

Jin, J. (1992). *The Finite Element Method in Electromagnetics*. John Wiley & Sons, New York (USA).

Levi, E. (1984). *Polyphase Motors – A Direct Approach to Their Design*. John Wiley & Sons, New York (USA).

Linke, M., Kennel, R., & Holtz, J. (2003). Sensorless speed and position control of synchronous machines using alternating carrier injection, in *Proc. of International Electric Machines and Drives Conference (IEMDC)*, Madison (USA), pp. 1211–1217.

Mademlis, C., Margaris, N., & Xypteras, J. (2000). Magnetic and thermal performance of a synchronous motor under loss minimization control, *IEEE Transaction on Energy Conversion*, Vol. 15 (No. 2): 135–142.

Mellor, P. H., Roberts, D., & Turner, D. R. (1991). Lumped parameter thermal model for electrical machines of TEFC design, *Electric Power Applications, IEE Proceedings B*, Vol. 138: 205–218.

Miller, T. (1989). *Brushless Permanent–Magnet and Reluctance Motor Drives*. Claredon Press – Oxford University Press, Oxford (UK).

Nakano, M., Kometani, H., & Kawamura, M. (2006). A study on eddy-current losses in rotors of surface permanent-magnet synchronous machines, *IEEE Transactions on Industry Applications*, Vol. 42 (No. 2): 429–435.

Ogasawara, S. & Akagi, H. (1998). An approach to real–time position estimation at zero and low speed for a PM motor based on saliency, *IEEE Transactions on Industry Applications*, Vol. 34 (No. 1): 163–168.

Salon, S. (1995). *Finite Element Analysis of Electrical Machine*, Kluwer Academic Publishers, USA.

Sanada, M., Hiramoto, K., Morimoto, S., & Takeda, Y. (2003). Torque ripple improvement for synchronous reluctance motor using asymmetric flux barrier arrangement, *IEEE Transactions on Industry Applications*, Vol. 40 (No. 4): 1076–1082.

Schuisky, W. (1967) *Berechnung Elektrischer Machinen*, Springer Verlag, Wien (Austria).

Shah, M. & Lee, S. B. (2006). Rapid analytical optimization of eddy-current shield thickness for associated loss minimization in electrical machines, *IEEE Transactions on Industry Applications*, Vol. 42 (No. 3): 642–649.

Shah, M. & Lee, S. (2009). Optimization of shield thickness of finite length rotors for eddy current loss minimization, *IEEE Transactions on Industry Applications*, Vol. 45 (No. 6): 1947–1953.

Slemon, G. R. & Straughen, A. (1980). *Electric Machines*, Addison–Wesley Pub. Co., New York (USA).

Vas, P. (1990). *Vector control of AC machines*, Claredon Press – Oxford Science Publications, Oxford (UK).

White, D. & Woodson, H. (1959). *Electromechanical Energy Conversion*, John Wiley and sons, New York (USA).

8

Finite Element Analysis of Desktop Machine Tools for Micromachining Applications

M. J. Jackson, L. J. Hyde, G. M. Robinson and W. Ahmed
Center for Advanced Manufacturing, College of Technology,
Purdue University, West Lafayette, IN,
USA

1. Introduction

The current interest in developing a manufacturing capability at the mixed scales is leading to a number of investigations concerned with the development of mesoscale machine tools (mMTs). The simulation of nanometric machining (Cook 1995, Inman 2001, Luo et al. 2003) and the effect of material microstructure (Komanduri et al. 2001 and Vogler et al. 2001) has led to the quest to construct machine tools capable of realizing 'bottom-up' fabrication processes in the general area of nanomanufacturing. The purpose of this paper is to investigate the use of a tetrahedral frame design to be used as a machine tool frame for meso, micro, and nanoscale machining applications. The problem with existing desk-top machine frames is the amount of vibration that is transmitted through the spindle, which affects the quality of surface finish and the dimensional accuracy imparted to the workpiece being machined. Owing to the way the spindle is mounted at the end of a cantilevered structure, low resonant frequencies can occur that are easily excited. In addition, the

Fig. 1. Tetrahedral machine tool frame.

amplitude of oscillation is more pronounced due to the geometry of the spindle mounting. An alternative approach is to design a vibration suppressing structure. When vibrations travel through a tetrahedral structure, they are canceled out or minimized due to the interference between the vibrating waves as they travel through the loops of the structure. The ability to minimize vibrations is needed because if the spindle oscillates during machining, an increase in the depth of cut will occur thus reducing the quality of surface finish, or dimensional accuracy of the machined part will be reduced significantly. Figure 1 shows the tetrahedral structure constructed for the purpose of this investigation. Modal analysis experiments were performed to investigate the structural response of the structure. Modal analysis experiments consisted of measuring the natural frequencies of the structure and deducing frequency response functions (F.R.F.) to determine the mode shapes of the structure. In addition, a finite element model (F.E.A.) model was constructed to compare to the experimental data, which also may be used for modeling any alterations to the design.

2. Analysis

The tetrahedral frame was initially analyzed from a numerical viewpoint using a closed-form solution and a numerical solution using finite element analysis.

2.1 Finite element model

Modal analysis of the tetrahedral structure using the finite element method was performed to obtain the natural frequencies and the mode shapes within the range of 0-8500 Hz, to compare to experimentally determined mode shapes. Modal analysis simulation was

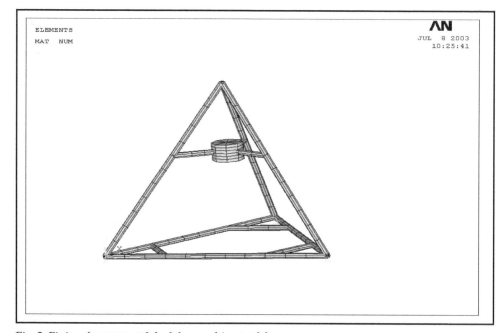

Fig. 2. Finite-element model of the machine tool frame.

performed using the finite element software package ANSYS 6.1. Model preparation was the first step in analyzing the modes of the tetrahedral structure (Stephenson 2002). This step involved creating a beam model of the structural members. The six bars that link the spheres and the reinforcement connections, which tie together the spindle sub-frame and the reinforcement bars were modeled using (ANSYS beam 188 elements), which have three translational degrees of freedom U_x, U_y, and, U_z at each node and three rotational degrees of freedom θ_x, θ_y and θ_z. The three rotational degrees of freedom were needed to accurately simulate the boundary conditions at the vertices of the structure. The finite element model, shown in Figure 2 consists of 115 elements and 513 nodes. The material properties of cold rolled steel were used in the modal analysis.

Each of the beam elements used enabled a geometric cross-section to be assigned. Each of the structural beams was given a circular cross-section of 0.75" diameter. The spindle holder was modeled by using a 3.5" O.D. 0.70" I.D. beam. This allowed the spindle holder to rotate and bend in a smooth manner. To simulate the spheres located at each of the vertices of the structure, a mass element (ANSYS mass 21) was used. The actual spheres of the structure were weighed and mass moments of inertia were calculated for them, and then input into the mass element model. Beam elements were chosen over solid elements to reduce the computation time required to solve the problem.

2.2 Closed-form solution model

Sample calculations were performed to approximate the dynamic response of the tetrahedral structure. The purpose of these calculations is to obtain a continuous model of the structure instead of a finite element approximation.

The structure was modeled as four spheres at each of the vertices of the tetrahedron, with springs simulating the structural links between them. Equation 1 was the equation used to generate a mathematical model of the structure.

$$[M]*\ddot{X}+[K]*X = 0 \tag{1}$$

Where, $[M]$ is the matrix of masses for each sphere, $\ddot{X}$ is the acceleration of each sphere, $[K]$ is the stiffness matrix for all of the structural links, and X is the displacement of each sphere. Equation 2 was used to model the stiffness (K) of each connecting rod only (axial displacements are considered in this formulation to decrease the complexity of the solution),

$$K = \frac{A*E}{L} \quad \text{where} \quad \begin{matrix} A = cross-sec\,tionalarea \\ E = ModulusofElasticity \\ L = LengthofRods \end{matrix} \tag{2}$$

Since this structure was modeled as a 9 degree-of-freedom (d.o.f.) system, the methods listed by Inman (2001) were used. This method assumes that each of the d.o.f.'s can be modeled by the superposition of several single d.o.f. systems. The structure is a three-dimensional structure, where each equation had to be related to a global coordinate system similar to the methods used in finite element formulations. There were three degrees of freedom for the top sphere and two degrees of freedom for the base spheres, which led to 9 possible natural frequencies. Damping was not considered in the mathematical modeling of this structure since it would create more difficulty in solving the equations.

2.2.1 Boundary conditions

The boundary conditions used for this structure had to allow the structure to translate along its base, but be constrained from movement in the vertical direction. To obtain these constraints, the top sphere was the only one that included a Z-axis component in the stiffness matrix, which allowed movement in the vertical direction. This allowed the vertical bars to move, which is required if the top sphere was allowed to oscillate. The base spheres did not include a vertical component since they were constrained from movement in the Z-direction. Since the displacement constraints existed only in the vertical direction, rigid body motion was to be expected, which led to three of the eigenvalues to be zero.

2.2.2 Free-body diagrams and derivation of equations of motion

Each of the spheres was modelled using a free-body diagram. The effect of gravity was neglected since it is taken into account by the equilibrium displacement of each of the structural links. The spring stiffness vectors corresponding to the direction illustrated the directions of displacement for the spheres. Damping was not considered in the mathematical modeling of this structure since it would create more difficulty in solving the equations and because it would not dramatically alter the natural frequencies of the structure. The equations of motion were derived from the free-body diagrams that were

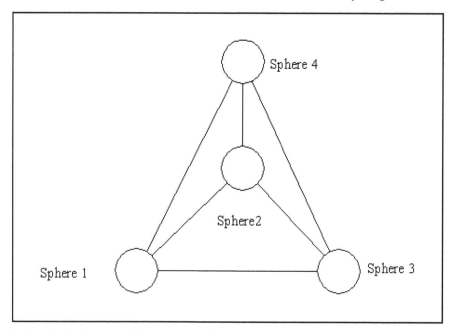

Fig. 3. Free-body diagram of the tetrahedral structure.

created for each of the spheres. There were two equations for each of the base spheres, and three equations for the top sphere. Since each of the links transmits oscillatory waves between the spheres, Newton's third law was used to simulate how each of the spheres responded to the force transmitted. Newton's third law of motion states that a body acted upon by a force will respond with an equal and opposite force to achieve equilibrium.

The first step to deriving a mathematical model was to relate the motion of each bar to a local X-Y-Z coordinate system. A brief example of how the equations for base sphere 1 were developed is shown. This example relates the motion for the links between sphere 1 and the spheres adjacent to it.

The free-body diagram shown in Figure 3 can be described using a set of local co-ordinate systems for each sphere connected to a global co-ordinate system connected by a series of rods. The system is shown in Figure 4.

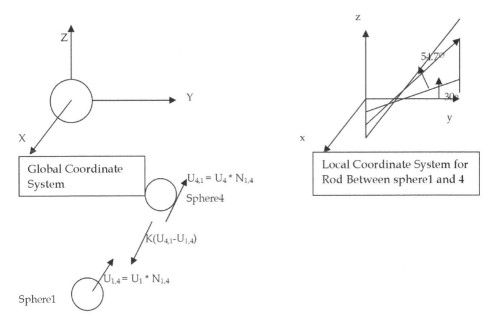

Fig. 4. Local and global co-ordinate system describing the relationship between spheres and connecting rods in the tetrahedral machine tool structure.

Figure 4 shows that,

$$N_{1,4} = e_{xy} \times \cos(54.7^{\circ}) - e_z \times \sin(54.7^{\circ}) \tag{3}$$

$$e_{xy} = e_y \times \cos(30^{\circ}) - e_x \times \sin(30^{\circ}) \tag{4}$$

Substituting for e_{xy} yields,

$$N_{4,1} = (e_y \times \cos(30^o) - e_x \times \sin(30^o)) \times \cos(54.7^o) - e_z \times \sin(54.7^o) \tag{5}$$

$$U_1 = (e_x x_1 + e_y y_1) \tag{6}$$

$$U_4 = (e_x x_4 + e_y y_4) \tag{7}$$

$$U_{1,4} = U_1 \bullet N_{1,4} \tag{8}$$

$$U_{4,1} = U_4 \bullet N_{1,4} \tag{9}$$

Equation 10 involves taking the dot product between the unit vector U_1 and $N_{1,4}$ to give,

$$U_{1,4} = (e_x \times x_1 + e_y \times y_1) \bullet ((e_y \times \cos(30^o) - e_x \times \sin(30^o)) \times \cos(54.7^o) + e_z \times \sin(54.7^o))$$

$$U_{1,4} = (-x_1 \times \sin(30^o) + y_1 \times \cos(30^o)) \times \cos(54.7^o) \tag{10}$$

Equation 11 involves taking the dot product between the unit vector U_4 and $N_{1,4}$ and reducing to give,

$$U_{4,1} = (e_x \times x_4 + e_y \times y_4 + e_z \times z_4) \bullet ((e_y \times \cos(30^o) - e_x \times \sin(30^o)) \times \cos(54.7^o) + e_z \times \sin(54.7^o))$$

That becomes,

$$U_{4,1} = (-x_4 \times \sin(30^o) + y_4 \times \cos(30^o)) \times \cos(54.7^o) + z_4 \times \sin(54.7^o) \tag{11}$$

Now writing the force balance,

$$F_{4,1} = K(U_{41} - U_{14}) \tag{12}$$

Thus Equation 12 becomes,

$$F_{4,1} = K[(-x_4 + x_1) \times \sin(30^o) \times \cos(54.7^o) + (y_4 - y_1) \times \cos(30^o) \times \cos(54.7^o) + z_4 \times \sin(54.7^o)] \tag{13}$$

The local equation of motion for the rod between sphere 1 and sphere 2 is shown in Figure 5. Figure 5 shows that,

$$N_{1,2} = -e_x \times \sin(60^o) + e_y \times \cos(60^o) \tag{16}$$

$$N_{2,1} = -e_x \times \sin(60^o) + e_y \times \cos(60^o) \tag{17}$$

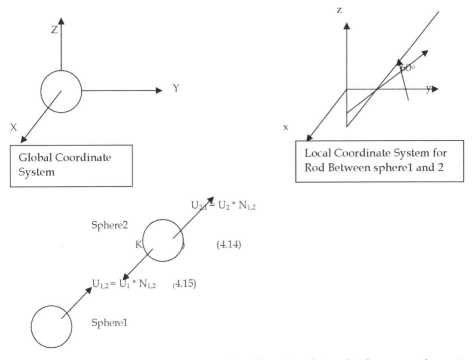

Fig. 5. Local and global co-ordinate system describing the relationship between spheres 1 and 2 and connecting rods in the tetrahedral machine tool structure.

$$U_1 = (e_x x_1 + e_y y_1) \tag{18}$$

$$U_2 = (e_x x_2 + e_y y_2) \tag{19}$$

$$U_{1,2} = U_1 \bullet N_{1,2} \tag{20}$$

$$U_{2,1} = U_2 \bullet N_{2,1} \tag{21}$$

$$U_{1,2} = (e_x \times x_1 + e_y \times y_1) \bullet (-e_x \times \sin(60^\circ) + e_y \times \cos(60^\circ)) \tag{22}$$

$$U_{2,1} = (e_x \times x_2 + e_y \times y_2) \bullet (-e_x \times \sin(60^\circ) + e_y \times \cos(60^\circ)) \tag{23}$$

Equation 23 reduces to,

$$U_{2,1} = (-x_2 \times \sin(60^\circ) + y_2 \times \cos(60^\circ)) \tag{24}$$

Now writing the force balance:

$$F_{2,1} = K(U_{2,1} - U_{1,2}) \tag{25}$$

Thus, Equation 25 becomes:

$$F_{2,1} = K[(-x_2 + x_1) \times \sin(60^\circ) + (y_2 - y_1) \times \cos(60^\circ)] \tag{26}$$

The local equation of motion for the rod between sphere 1 and sphere 3 is shown in Figure 6.

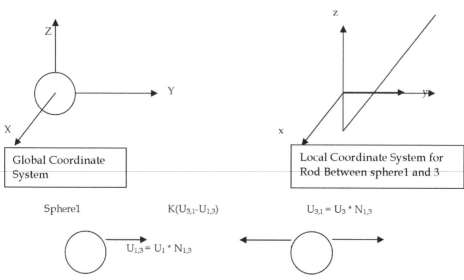

Fig. 6. Local and global co-ordinate system describing the relationship between spheres 1 and 3 and connecting rods in the tetrahedral machine tool structure.

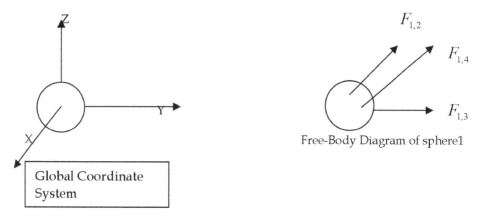

Fig. 7. Global co-ordinate system representing the tetrahedral machine tool structure in the form of a rod and sphere free-body diagram.

Now that the local equations of motion have been derived, the global system of equations must be defined. Summing forces in the X, and Y directions accomplish this task. The Z-direction is not considered since the sphere is restricted from movement in this direction. Figure 7 shows the arrangement for the definition of the global co-ordinates of rods and spheres.

$$F_{3,1} = K(y_3 - y_1) \tag{27}$$

$$\sum F_{x1} = -F_{1,2} \times \cos(30^o) - F_{1,4} \times \cos(54.7^o) \times \sin(30^o) = M_1 \times \ddot{x}_1 \tag{28}$$

Substituting equations Equation 13, and Equation 26 yields,

$$-K \times \cos(30^o) \times [(-x_2 + x_1) \times \sin(60^o) + (y_2 - y_1) \times \cos(60^o)]$$
$$-K \times \cos(54.7^o) \times \sin(30^o) \times [(-x_4 + x_1) \times \sin(30^o) \times \cos(54.7^o) + \tag{29}$$
$$(y_4 - y_1) \times \cos(30^o) \times \cos(54.7^o) + z_4 \times \sin(54.7^o)] = M_1 \times \ddot{x}_1$$

D'Alembert's Principle is used so that the equation will be in the correct form for substituting into the stiffness matrix, K, and into the mass matrix, M, thus,

$$K \times \cos(30^o) \times [(-x_2 + x_1) \times \sin(60^o) + (y_2 - y_1) \times \cos(60^o)]$$
$$K \times \cos(54.7^o) \times \sin(30^o) \times [(-x_4 + x_1) \times \sin(30^o) \times \cos(54.7^o) + \tag{30}$$
$$(y_4 - y_1) \times \cos(30^o) \times \cos(54.7^o) + z_4 \times \sin(54.7^o)] + M_1 \times \ddot{x}_1 = 0$$

Summing forces in the Y-direction gives,

$$\sum F_{y1} = F_{1,2} \times \cos(60^o) + F_{1,3} + F_{1,4} \times \cos(54.7^o) \times \cos(30^o) = M_1 \times \ddot{y}_1 \tag{31}$$

Substituting equations Equation 13, and Equations 26 and 27 yields,

$$K \times \cos(60^o) \times [(-x_2 + x_1) \times \sin(60^o) + (y_2 - y_1) \times \cos(60^o)]$$
$$+K \times [(y_3 - y_1)]$$
$$+K \times \cos(54.7^o) \times \cos(30^o) \times [(-x_4 + x_1) \times \sin(30^o) \times \cos(54.7^o) + \tag{32}$$
$$(y_4 - y_1) \times \cos(30^o) \times \cos(54.7^o) + z_4 \times \sin(54.7^o)] = M_1 \times \ddot{y}_1$$

Again, D'Alembert's Principle is used so that the equation will be in the correct form for substituting into the stiffness matrix, K, and the mass matrix, M,

$$-K \times \cos(60^o) \times [(-x_2 + x_1) \times \sin(60^o) + (y_2 - y_1) \times \cos(60^o)]$$
$$-K \times [(y_3 - y_1)]$$
$$-K \times \cos(54.7^o) \times \cos(30^o) \times [(-x_4 + x_1) \times \sin(30^o) \times \cos(54.7^o) + \tag{33}$$
$$(y_4 - y_1) \times \cos(30^o) \times \cos(54.7^o) + z_4 \times \sin(54.7^o)] + M_1 \times \ddot{y}_1 = 0$$

Equations 30 and 33 may be substituted into the stiffness matrix K and the mass matrix M.

2.2.3 Solution of matrices

After the equations for the 9 degrees-of-freedom were written, they were solved to find the natural frequencies of vibration. Two matrices were built, one for the mass matrix $[M]$, and a stiffness matrix $[K]$. Once the matrices were built the eigen function was used to solve the eigenvalues and eigenvectors. Once the eigenvalues were solved, the square root was taken followed by dividing by 2π, which yielded the natural frequencies of the system in units of Hertz. There were three zero eigenvalues, which was expected since rigid-body motion in the X and Y directions and rotation in the XY plane was allowed. There were also two repeated eigenvalues that occurred due to symmetry of the structure. This was considered to be trivial since the eigenvectors would not be used for generating mode shapes. A list of the natural frequencies found from the mathematical model is shown in Table 1.

Mode set	Frequency (Hertz)
0	0
1	0
2	0
3	0
4	19.6
5	19.6
6	33.0
7	34.6
8	34.6
9	50.2

Table 1. Natural frequencies generated for the tetrahedral machine tool structure based on the results using the closed-form model.

2.3 Model verification

A comparison of the closed-form solution and the simplified finite-element model of the structure was performed. This was conducted to determine whether the finite element software, ANSYS, would yield accurate results that would be comparable to the mathematical model. The bar model from ANSYS consists of 6 bar elements and 4 mass elements, which are located at the vertices of the structure. The bar element was a beam 3 element, which only allows movement along its axis, thus tension or compression. This was used since the mathematical model was formulated in this manner. The structure was constrained from movement in the U_y direction and rotations about the Z-axis, and the X axis. The model and the constraints are shown in Figure 8.

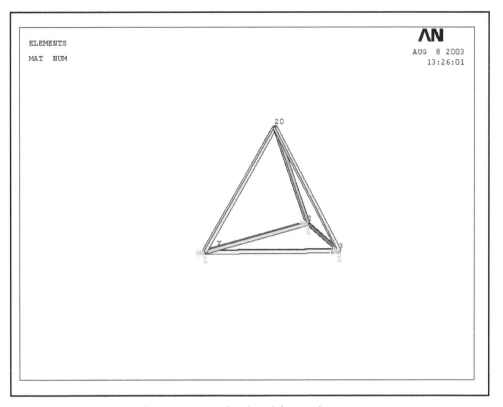

Fig. 8. A simple bar model to compare to the closed-form solution.

Finite element bar model (ANSYS)		Closed-form solution model		% Difference
Mode set	Frequency	Mode set	Frequency	
	(Hz)		(Hz)	
1	0	1	0	0%
2	0	2	0	0%
3	0	3	0	0%
4	23.5	4	19.6	-20%
	Not available	5	19.6	0%
	Not available	6	33.0	0%
5	37.1	7	34.6	-7%
	Not available	8	34.6	0%
6	51.4	9	50.2	-2%

Table 2. Comparison of finite element bar model to the closed-form solution.

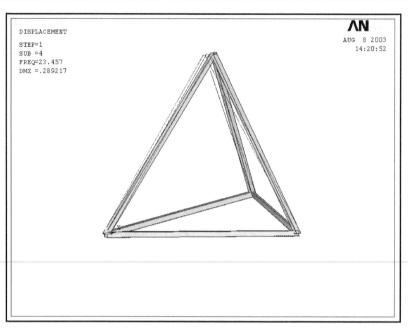

Fig. 9. Finite-element bar model showing displacement of the structure at a frequency of 23.5 Hz.

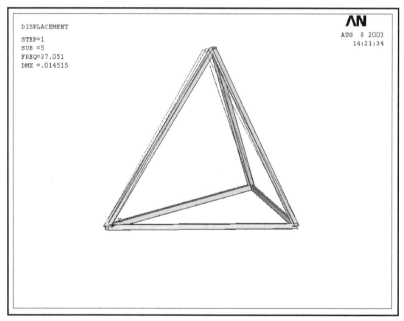

Fig. 10. Finite element bar model showing displacement of the structure at a frequency of 37 Hz.

Table 2 lists the frequencies found from the finite element model and compares them to the closed-form solution. Both models show three rigid body modes and have three frequencies that are similar. Only six frequencies were found from ANSYS since it only had six degrees of freedom, one for each of the bar elements. In addition, it can be seen as the frequencies increase the percentage difference decreases. The finite element beam model built to compare with the measured model was not used to compare with the closed-form solution because it did not include any bending modes, therefore the two do not correlate. However the closed-form solution is useful, since it proved that ANSYS compares accurately using a simplified bar model, and should compare well to the measured model.

Since the finite element model compares reasonably well with the closed-form solution, the mode shapes that were generated from ANSYS are shown in Figures 9-11.

It can be shown from Figure 9 and Figure 10 that the top vertex oscillates along one of the rods. One of the upper rods shortens, whereas the opposing rods rotate about the base vertices.

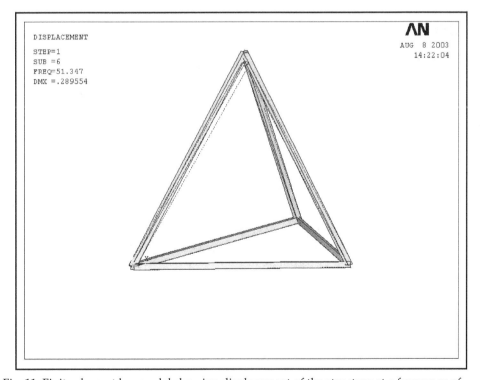

Fig. 11. Finite element bar model showing displacement of the structure at a frequency of 51.3 Hz.

Figure 11. Illustrates how the structure oscillates at 51.3 Hz. It can be seen that each of the upper rods elongates thus causing the top vertex to increase its height.

3. Experimental

The impact hammer test has become a widely used device for determining mode shapes. The peak impact force is nearly proportional to the mass of the head of the hammer and its impact velocity. The load cell in the head of the hammer provides a measure of the impact force. This data is used to compute the frequency response function (F.R.F.). The use of an impact hammer avoids the mass-loading problem and is much faster to use than a shaker. An impact hammer consists of a hammer with a force transducer built into the head of the hammer. The hammer is used to impart an impact to the structure and excite a broad range of frequencies. The impact event is supposed to approximate a Dirac-delta function [2].

3.1 Experimental method

The tetrahedral structure [6] was placed on a granite table in order to gain accelerometer measurements, thus the structure was allowed to freely move longitudinally and transversely across the table. The roving accelerometer approach was used for all of the measurements. The center of the spindle frame was used as the excitation point for the structure. The accelerometer was placed at various points of interest about the structure.

3.2 Experimental procedure

The data acquisition system was set up to take data at a sampling frequency, F_s, of 17000 Hz for 8192 points with a delay of 100 points. The voltage range on both channels was set to +\-5 volts. Data was acquired in the time domain by averaging 8 ensembles and storing the data in binary format for use by Matlab software. This data was used to find the natural frequencies of the structure and their corresponding mode shapes. While applying the roving accelerometer technique, the structure was excited in the center of the spindle sub-frame and data was acquired at points 1-28. Before the time domain data was stored, it was filtered to remove any aliasing that might have occurred from under-sampling. This was accomplished by installing an analog filter between the power supply and the P.C. The frequency was set at 8500 Hz, which corresponds to the Nyquist frequency of the measured data. After the data was recorded, it was translated into a binary file to be fed into Matlab software. The method used on the F.R.F. data of a multi-degree-of-freedom structure is the single-degree-of-freedom-curve-fit (S.D.O.F.). In this method the frequency response function for the compliance is sectioned off into frequency ranges bracketing each successive peak. Each peak is analyzed by assuming that it is the F.R.F. of a single-degree-of-freedom system. This assumes that in the vicinity of resonance the F.R.F. is dominated by that single mode. Once the frequency response function (F.R.F.) is completed for the chosen data points of a structure, it is then appropriate to compute the natural frequencies, damping ratios and modal amplitudes with each resonant peak. An example of one of the F.R.F.'s is shown in Figure 13. The damping ratio associated with each peak is assumed to be the modal damping ratio Zeta, ξ. The modal damping ratio Zeta is related to the frequencies corresponding to Equation 34.

$$|H(\omega_a)| = |H(\omega_b)| = \frac{|H(\omega_d)|}{\sqrt{2}} \tag{34}$$

And $\omega_b - \omega_a = 2\zeta\omega_d$, so that

$$\zeta = \frac{\omega_b - \omega_a}{2\omega_d} \tag{35}$$

ω_d is the damped natural frequency at resonance such that ω_a and ω_b satisfy Equation 3. The condition of Equation 34 is termed the 3 dB down point. Both the natural frequency and the damping ratio Zeta may be found using this method. Once the values of ω_a and ω_b are determined, then ζ is found for the structure at the prescribed frequency (Equation 35). This method was used in the software to experimentally determine the damping and mode shapes. Figure 13 gives an example of F.R.F. data set that was found from the tetrahedral structure.

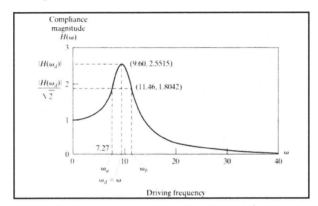

Fig. 12. Magnitude of the frequency response function, illustrating the calculation of the modal damping ratio by using the quadrature peak-picking method for lightly damped systems.

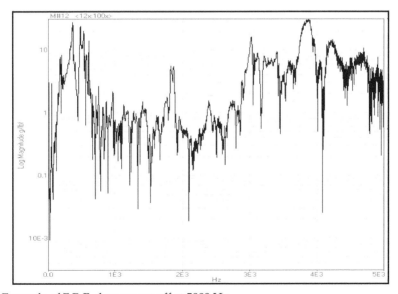

Fig. 13. Example of F.R.F. data set cut off at 5000 Hz.

3.3 Experimental analysis

Using the measured data obtained from Me-Scope software, a model was constructed and the data was used to find structural damping and mode shapes. At first it was thought that the data was too low since the operating frequencies of the spindle are above 4500 Hz. However the operating frequencies of the spindle could excite lower frequencies while machining. Therefore, this data is useful if the structure is excited at these frequencies by some other means, such as localized impacts the structure might experience during a machining operation. This is shown in the following series of illustrations at the chosen frequencies (Figures 15 –20). The measured data for each node was adjusted such that the axis of orientation corresponded with the orientation of the accelerometer.

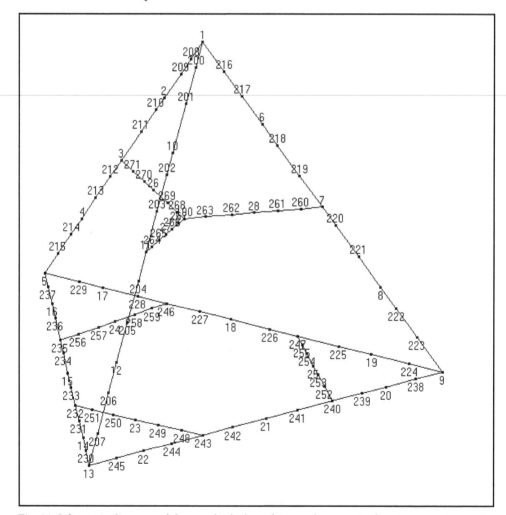

Fig. 14. Schematic diagram of the tetrahedral machine tool structure showing experimental points-of-measurement for determining mode shapes.

Figure 14 illustrates where mode shape measurements were taken during the experimental phase of this study (numbers 1-28 represent actual data points, whereas the other numbers were used for interpolation between measurement points.

4. Results and discussion

The measured data compared accurately with the finite element results. It was found that the position of the centre of the spindle proved to be a point inside the structure that experienced minimal oscillations. It appeared that the structure was kinematically balanced such that different parts of the structure had oscillations that were out of phase with other parts. The tetrahedral structure was analyzed in its working orientation. The results are tabulated in Table 1.

Measured data	Finite element results	
(Me-Scope software)	(ANSYS software)	% Difference
125	125	0%
203	200	1%
401	407	-1%
534	535	0%
601	600	0%
1070	1085	-1%
1820	1794	1%

Table 3. Comparison between measured and finite element calculations.

Not all of the results are listed, only those of interest. The first column is the measured natural frequency, followed by the finite element generated natural frequency in the second column. The observed modes of interest are shown in Figures 15-17. The following figures illustrate how the tetrahedral structure oscillates at various frequencies. The measured data mode shape is given first, followed by a corresponding finite element generated mode shape. As the frequency is increased, the results from the finite element model seem to diverge from the measured mode shapes. It is thought that as oscillation modes increase they tend to depart from Bernoulli beam theory upon which the finite element generated results depend. For most of the natural frequencies, the amount of oscillation of the spindle is small, or approximately zero, which is preferred since the amount of spindle oscillation from equilibrium is translated directly to the machined workpiece. The results from ANSYS above 1794 Hz did not coincide with what was measured, therefore no comparison was made. However, measured frequencies above 1820 Hz are shown because they are useful for future design revisions to the structure. Axial responses, as well as transverse responses, from the measured data were used to compare to the finite element results. Torsional data was ignored since it was not recorded using the accelerometer and the Me-Scope measurement software. It can be seen from the percent difference that the results from ANSYS have a natural frequency that resembles the measured results. However, they do not converge exactly instead the results oscillate about the measured data.

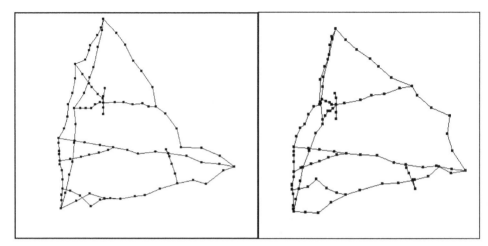

Fig. 15a. Me-Scope measured mode shape data (i.e., displacement) at a frequency of 125 Hz.

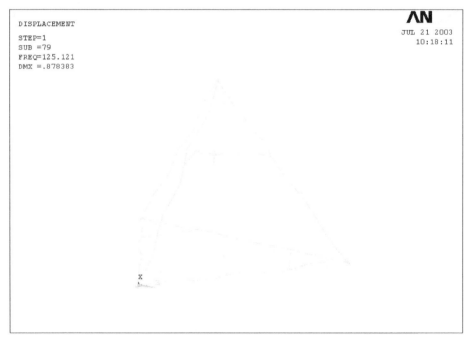

Fig. 15b. Finite element generated model of mode shape data (displacement) at a frequency of 125 Hz.

The measured mode shape data using ME-Scope software at a frequency of 125 Hz shows an axial deflection for the spindle frame. However, the spindle itself remains stationary. The ANSYS model shows bending in the spindle sub-frame. Both finite element models show axial bending modes in the structural bars.

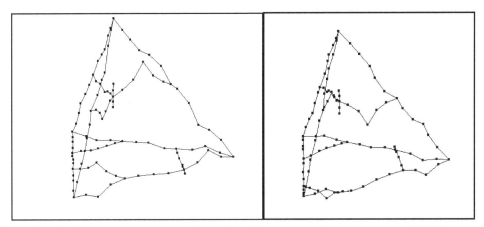

Fig. 16a. Me-Scope measured mode shape data (displacement) at a frequency of 232 Hz.

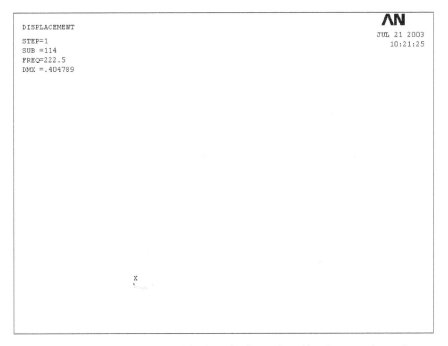

Fig. 16b. Finite element generated model of mode shape data (displacement) at a frequency of 222 Hz.

The measured mode shape at 232 Hz illustrates how the structure cancels out oscillations that are transmitted through the spindle. It is apparent from the measured mode shape as well as the finite element model, how various structural members are out of phase, which prevents any displacement of the spindle from its equilibrium position thus achieving a preferred effect for machining.

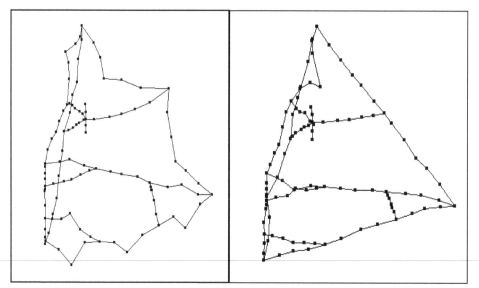

Fig. 17a. Me-Scope measured mode shape data (displacement) at a frequency of 1820 Hz.

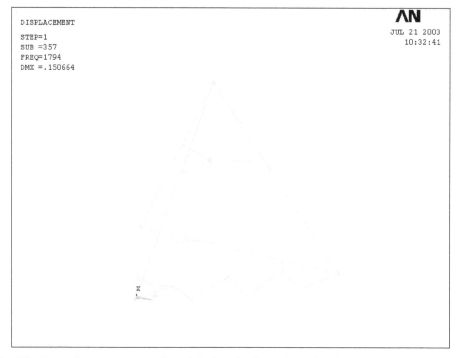

Fig. 17b. Finite element generated model of mode shape data (displacement) at a frequency of 1794 Hz.

The measured data vaguely coincides with the finite element model once the frequencies reach approximately 1800 Hz, as illustrated from the images from Me-Scope at 1820 Hz and the finite element results at 1794 Hz. For this reason the finite element results have been omitted above 1820 Hz. This may be due to inadequate modeling of the structural connections, but most likely due to Bernoulli beam theory not being applicable at these frequencies. The only characteristic that is common to both of the models is the restricted oscillation of the spindle. It can readily be seen from the following Figures 18 – 20 that there is virtually zero oscillation in the spindle at most of the measured frequencies, this is accompanied by the finite element model as well. The reason for omitting results above 4460 Hz is because the F.R.F. from the measured data was not clean, thereby resembling noise, which is not useful for an adequate conclusion to be made. This is because the impact hammer method of exciting a structure is limited to approximately 4000 Hz.

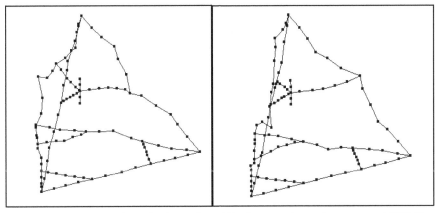

Fig. 18. Me-Scope measured mode shape data (displacement) at a frequency of 2890 Hz.

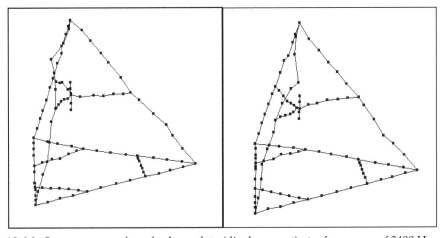

Fig. 19. Me-Scope measured mode shape data (displacement) at a frequency of 3400 Hz.

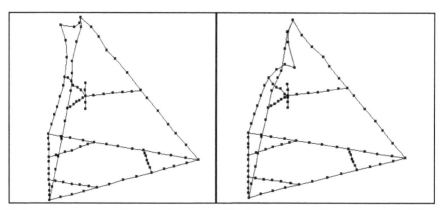

Fig. 20. Me-Scope measured mode shape data (displacement) at a frequency of 4460 Hz.

5. Conclusions

It is concluded that the finite element model prediction compares well with measured data at low frequencies. Owing to this fact, the finite element model may be used for future design improvements to the structure. It can be seen from the experimental and the measurement results that multiple constraints on the spindle enhance the ability of the structure to resist excitation. One possible reason for the structure's oscillation is probably due to the lack of passive damping. Therefore, it is recommended that improvements be made to improve passive damping of oscillations.

6. Acknowledgements

The author thanks Inderscience for use of material that was published as: Dynamic response of a tetrahedral nanomachining machine tool structure by Mark J. Jackson, Luke J. Hyde, Grant M. Robinson, Waqar Ahmed DOI: 10.1504/IJNM.2006.011378, International Journal of Nanomanufacturing, 2006, Volume, Number 1, p.p. 26-46. Full copyright is retained by Inderscience.

7. References

Cook, R. D., *Finite Element Modeling For Stress Analysis*, John Wiley & Sons Inc., New York, 1995.

Inman, D. J., *Engineering Vibration*, Prentice Hall, Upper Saddle River, New Jersey, 2001

Komanduri, R., Chandrasekaran, N., and Raff, L. M., 2001, *Molecular dynamics simulation of the nanometric cutting of silicon*, Philosophical Magazine, B81, 1989-2019.

Luo, X., Cheng, K., Guo, X., and Holt. R., 2003, *An investigation into the mechanics of nanometric cutting and the development of its test bed*, Int. J. Prod. Res., 41, 1449-1465.

Stephenson, D.J., et al: *"Ultra Precision Grinding Using the Tetraform Concept"*, Abrasives Magazine, February/March 2002, p.p.12-16.

Vogler, M. E., De Vor, R. E., and Kapoor, S. G., 2001, *Microstructure – level force prediction model for micro-milling of multi-phase materials*, Proceedings of the International Mechanical Engineering Conference and Exposition, A.S.M.E. Manufacturing Engineering Division, 12, p.p., 3 – 10.

9

Semi-Analytical Finite Element Analysis of the Influence of Axial Loads on Elastic Waveguides

Philip W. Loveday[1],
Craig S. Long[1] and Paul D. Wilcox[2]
[1]*CSIR Material Science and Manufacturing,*
[2]*University of Bristol,*
[1]*South Africa*
[2]*United Kingdom*

1. Introduction

Guided wave ultrasound is a promising technology for non-destructive inspection and monitoring of long slender structures such as pipes and rails. These structures are effectively one-dimensional waveguides and therefore a large length of the structure can be inspected from a single transducer location. A good introduction to guided wave inspection is available in (Rose, 2002). During the design of guided wave inspection systems it is advantageous to understand the modes of propagation, to predict how these modes interact with the damage to be detected and to be able to transmit and receive these modes independently (Lowe et al., 1998). Analytical solutions for wave propagation in circular cylinders are well known (Gazis, 1959) and may be used when developing pipe monitoring systems. When the cross-sectional geometry is more complex, such as in the case of rails, it becomes necessary to employ numerical solutions.

It is possible to compute the wavenumber – frequency relations of propagating modes using conventional three – dimensional continuum finite element models, with appropriate boundary conditions, as was demonstrated by Thompson (1997). The advantage of this method is that it can be implemented using commercially available finite element codes, but requires significant user post processing, and the solution of many different model lengths. Alternately, semi-analytical finite elements (SAFE) can be specially formulated to efficiently analyse these problems. The formulation of these elements includes complex exponential functions to describe the wave propagation along the waveguide and conventional finite element interpolation functions across the cross-section. Therefore, only a two – dimensional mesh of the cross-section of the waveguide is required, which results in a significant reduction in the amount of computation required. These elements are sometimes called waveguide finite elements and have been implemented by a number of research groups (Aalami, 1973; Lagasse, 1973; Gavrić, 1995; Hayashi et al., 2003; Damljanović & Weaver, 2004a; Bartoli et al., 2006; Predoi Mihai et al., 2007; Castaings & Lowe, 2008 and Ryue et al.,

2008). In addition to computing dispersion properties of complex geometry waveguides, the SAFE approach has been applied to a number of other problems, including forced response computations (Damljanović & Weaver, 2004b), modelling of piezoelectric transducers attached to waveguides (Loveday, 2008), determining the power input to waveguides (Nilsson & Finnveden, 2007), and analysing the scattering of waves at discontinuities (Baronian et al., 2009). Authors use slightly different formulations and two popular formulations will be compared in Section 2.

It is well known that the application of an axial load to a beam can influence the natural frequencies of the beam. Similarly, the application of an axial load to a waveguide can influence the wave propagation characteristics of the waveguide. In the case of continuously welded rail, the axial load in the rail is an important parameter. These rails are installed to generally be in tension but the amount of tension depends on the temperature. If the rail goes into compression there is a danger of buckling, which can lead to derailments while if the tension is excessive, the probability of fatigue cracks is increased and this can lead to rail breaks and derailments. A guided wave system was developed to continuously monitor rails for breaks (Loveday, 2000) and it would be a valuable addition if the same system could monitor the rail for compression in the rail before buckling. There has been considerable research into the possibility of using guided waves to measure the axial load in rails. Low frequency flexural waves were investigated by Damljanović & Weaver (2005) who proposed to use a scanning laser vibrometer to measure displacements of points along the rail and then to extract the wavenumber of the flexural wave at 200Hz (Damljanović & Weaver, 2004b). This method requires that the rail be released from the sleepers for a considerable length. Chen & Wilcox (2006) investigated the use of higher frequency guided waves for measuring loads in rods. Simulation results clearly demonstrated that the phase velocity and group velocity are sensitive to changes in load.

The influence of the axial load on wave propagation characteristics may be analysed by three – dimensional finite element models, although the process is tedious and difficult at higher frequencies where numerous waves propagate (Chen & Wilcox, 2007). Recently the SAFE method was extended to include axial loads (Loveday, 2009) and it was demonstrated that the required modification to an existing SAFE code is trivial. This extension is described in Section 3 of this chapter. The use of SAFE to analyse the influence of axial loads on wave propagation in rails offers the advantage of computational efficiency. In addition because of the analytical nature of the method, it is possible to directly compute certain sensitivities that would otherwise have to be computed by a finite difference method (Loveday & Wilcox, 2010). These possibilities are described in Section 4 while numerical results are presented in Section 5.

2. Semi-analytical finite element formulation for elastic waveguides

The formulation presented here follows that of Gavrić (1995) who choose the complex exponential functions along the waveguide to have the axial terms phase shifted by $\pi/2$ relative to the in-plane terms. This choice leads directly to symmetric matrices and is described in Section 2.1. Many of the other SAFE formulations used do not result in a symmetric stiffness matrix, which is inconvenient for eigensolvers. These formulations will not be considered here. A different approach was adopted by Damljanović & Weaver (2004a) who used a transformation to obtain symmetric matrices. These matrices are the same as those obtained by Gavrić except that the sign of one matrix is reversed. A comparison of the two approaches is outlined in Section 2.2.

2.1 Formulation by Gavrić

The displacement field in an elastic waveguide, extending in the z direction, may be written as a complex exponential along the waveguide and a finite element approximation over the cross-section. The displacement fields (u, v, w) employed by are Gavrić (1995) were chosen to take the form:

$$
\begin{aligned}
u(x,y,z,t) &= u(x,y) \cdot e^{-j(\kappa z - \omega t)} \\
v(x,y,z,t) &= v(x,y) \cdot e^{-j(\kappa z - \omega t)} \\
w(x,y,z,t) &= w(x,y) \cdot e^{-j(\kappa z - \omega t - \pi/2)} = j \cdot w(x,y) \cdot e^{-j(\kappa z - \omega t)}
\end{aligned}
\tag{1}
$$

where, z is the coordinate in the direction along the waveguide, κ the wavenumber and ω the frequency. $u(x,y)$, $v(x,y)$ and $w(x,y)$ are the interpolated displacements in the x, y and z directions respectively. The strain energy of an infinitesimal element of the waveguide is,

$$
s = \frac{1}{2}\varepsilon^* \cdot c \cdot \varepsilon = \frac{1}{2}\left[\kappa^2 \cdot s_2 + \kappa \cdot s_1 + s_0\right],
\tag{2}
$$

where, * denotes complex transpose, k_0, k_1 and k_2 are defined below, ε and c are the strain and elasticity matrices defined in standard form as:

$$
\varepsilon(x,y,z,t) = \left\{
\begin{array}{c}
\dfrac{\partial u}{\partial x} \\[4pt]
\dfrac{\partial v}{\partial y} \\[4pt]
\dfrac{\partial w}{\partial z} \\[4pt]
\dfrac{\partial u}{\partial y} + \dfrac{\partial v}{\partial x} \\[4pt]
\dfrac{\partial v}{\partial z} + \dfrac{\partial w}{\partial y} \\[4pt]
\dfrac{\partial u}{\partial z} + \dfrac{\partial w}{\partial x}
\end{array}
\right\};
\quad
c = \frac{E}{(1+v)(1-2v)}
\begin{bmatrix}
1-v & v & v & & & \\
v & 1-v & v & & & \\
v & v & 1-v & & & \\
& & & \frac{1-2v}{2} & & \\
& & & & \frac{1-2v}{2} & \\
& & & & & \frac{1-2v}{2}
\end{bmatrix}
\tag{3}
$$

where v is Poisson's ratio and E is elastic modulus. The strains due to the displacement field in (1) are separated into a function over the cross-section, multiplied by the complex exponential. Furthermore, for convenience the terms that are dependent on the wavenumber are separated from those independent of wavenumber. The result is:

$$\varepsilon(x,y,z,t) = \varepsilon(x,y) \cdot e^{-j(\kappa z - \omega t)}$$

$$\varepsilon(x,y) = \varepsilon_0(x,y) + \kappa \varepsilon_1(x,y)$$

$$\varepsilon(x,y) = \begin{Bmatrix} \dfrac{\partial u}{\partial x} \\ \dfrac{\partial v}{\partial y} \\ \kappa w \\ \dfrac{\partial u}{\partial y} + \dfrac{\partial v}{\partial x} \\ -j\kappa v + j\dfrac{\partial w}{\partial y} \\ -j\kappa u + j\dfrac{\partial w}{\partial x} \end{Bmatrix} ; \quad \varepsilon_0(x,y) = \begin{Bmatrix} \dfrac{\partial u}{\partial x} \\ \dfrac{\partial v}{\partial y} \\ 0 \\ \dfrac{\partial u}{\partial y} + \dfrac{\partial v}{\partial x} \\ j\dfrac{\partial w}{\partial y} \\ j\dfrac{\partial w}{\partial x} \end{Bmatrix} ; \quad \varepsilon_1(x,y) = \begin{Bmatrix} 0 \\ 0 \\ w \\ 0 \\ -jv \\ -ju \end{Bmatrix} \tag{4}$$

The terms in the strain energy, which are independent, linearly and quadratically dependent can now be expressed as follows:

$$s_0 = \varepsilon_0{}^* c \varepsilon_0$$
$$s_1 = \varepsilon_0{}^* c \varepsilon_1 + \varepsilon_1{}^* c \varepsilon_0 . \tag{5}$$
$$s_2 = \varepsilon_1{}^* c \varepsilon_1$$

Four-noded isoparametric elements were used in this research although quadratic elements are more efficient (Andhavarapu et al., 2010). The standard interpolation matrix N, which is a function of local coordinate (ξ, η), relates the nodal degrees of freedom to the coordinate and displacement distributions as follows:

$$\bar{u}(\xi, \eta) = N(\xi, \eta)u ; \quad u = [u_1 \; v_1 \; w_1 \; u_2 \; v_2 \; w_2 \; u_3 \; v_3 \; w_3 \; u_4 \; v_4 \; w_4]^T . \tag{6}$$

Where u is the vector of nodal displacements. The strains may now be written,

$$\varepsilon_0(\xi, \eta) = B_0(\xi, \eta)u; \quad \varepsilon_1(\xi, \eta) = B_1(\xi, \eta)u , \tag{7}$$

and the strains $\varepsilon_0(\xi, \eta)$, $\varepsilon_1(\xi, \eta)$ are related to $\varepsilon_0(x,y)$, $\varepsilon_1(x,y)$ by the Jacobian , J, as usual. Integration is performed over the area of the elements to give the elemental matrices.

$$k_0 = \int B_0 c B_0{}^* \det J \; d\xi d\eta$$
$$k_1 = \int \left[B_0 c B_1{}^* + B_1 c B_0{}^* \right] \det J \; d\xi d\eta$$
$$k_2 = \int B_1 c B_1{}^* \det J \; d\xi d\eta$$
$$m = \int N \rho N^* \det J \; d\xi d\eta \tag{8}$$

The element matrices are assembled into the system equations of motion. For free vibration the equation of motion is:

$$M\ddot{U} + \left[\kappa^2 \cdot K_2 + \kappa \cdot K_1 + K_0 \right] U = 0 \, , \tag{9}$$

where capitals indicate the assembled counterpart of the lowercase elemental matrices. It should be noted that the mass (M) and stiffness (K_0, K_1, K_2) matrices are all real and symmetric.

2.2 Comparison with formulation by Damljanović and Weaver

Damljanović & Weaver (2004a) employed the following functions to describe the displacement field:

$$
\begin{aligned}
u(x,y,z,t) &= u(x,y) \cdot e^{-j(\kappa z - \omega t)} \\
v(x,y,z,t) &= v(x,y) \cdot e^{-j(\kappa z - \omega t)} \\
w(x,y,z,t) &= w(x,y) \cdot e^{-j(\kappa z - \omega t)}
\end{aligned}
\tag{10}
$$

After applying a procedure similar to that in Section 2.1 they obtain the system of equations of motion:

$$M\ddot{U} + \left[\kappa^2 \cdot K_2 + j\kappa \cdot \tilde{K}_1 + K_0 \right] U = 0 \tag{11}$$

In this case the matrix $\tilde{K}_1$ is skew symmetric and they use the transformation matrix T to transform this matrix to a symmetric matrix. The matrix T has the form,

$$
T = \begin{bmatrix}
1 & & & & & & \\
 & 1 & & & & & \\
 & & j & & & & \\
 & & & 1 & & & \\
 & & & & 1 & & \\
 & & & & & j & \\
 & & & & & & \ddots
\end{bmatrix}
\tag{12}
$$

and can be thought of as a transformation from the physical coordinates to a transformed coordinate system,

$$U_t = TU_p \tag{13}$$

where, U_p are the displacements in the physical coordinates and U_t are the displacements in the transformed coordinates. The transformation has the properties $T^*T = TT^* = I$ and the reverse transformation is $U_p = T^*U_t$. Applying this transformation to the matrices in (11) has the following result:

$$TK_2T^* = K_2$$
$$TK_0T^* = K_0$$
$$TMT^* = M \tag{14}$$
$$T\tilde{K}_1T^* = -j\hat{K}_1$$

The transformation has no effect on the symmetric matrices, but the skew symmetric matrix is transformed to a symmetric matrix. The equations of motion in the transformed coordinates are then,

$$M\ddot{U}_t + \left[\kappa^2 \cdot K_2 + \kappa \cdot \hat{K}_1 + K_0\right]U_t = 0 \tag{15}$$

In order to compare the two formulations it is noted that the displacement functions used by Gavrić (1) may be written as,

$$U_p = TU_t \tag{16}$$

which is the conjugate of the transformation used by Damljanović & Weaver which was $U_p = T^*U_t$. The result of the different transformations is that, in the transformed coordinates, the matrix K_1 in Garvić's method is equal to $-\hat{K}_1$ in Damljanović & Weaver's formulation. The other matrices are the same.

If one were to start with the functions:

$$u(x,y,z,t) = u(x,y) \cdot e^{-j(\kappa z - \omega t)}$$
$$v(x,y,z,t) = v(x,y) \cdot e^{-j(\kappa z - \omega t)}$$
$$w(x,y,z,t) = -j \cdot w(x,y) \cdot e^{-j(\kappa z - \omega t)} \tag{17}$$

and follow Garvić's method, the transformed matrices of Damljanović & Weaver would be obtained directly.

3. Extension of SAFE to include axial load

The presence of an initial stress or load introduces additional terms in the strain energy which therefore lead to additions to the stiffness matrix. The SAFE method is extended to include axial loads in one-dimensional waveguides in this section.

When analysing small amplitude elastic waves we generally use the linear infinitesimal strain – displacement relations. However, when initial finite strains are present in the structure, it is necessary to make use of the full strain-displacement relation (Rose, 1999). The linear (ε) and full (E) strain - displacement relationships may be written as,

$$
\varepsilon = \left\{ \begin{array}{c} \dfrac{\partial u}{\partial x} \\[2mm] \dfrac{\partial v}{\partial y} \\[2mm] \dfrac{\partial w}{\partial z} \\[2mm] \dfrac{\partial u}{\partial y} + \dfrac{\partial v}{\partial x} \\[2mm] \dfrac{\partial v}{\partial z} + \dfrac{\partial w}{\partial y} \\[2mm] \dfrac{\partial u}{\partial z} + \dfrac{\partial w}{\partial x} \end{array} \right\} ; \quad E = \varepsilon + \left\{ \begin{array}{c} \dfrac{1}{2}\left[\left(\dfrac{\partial u}{\partial x}\right)^2 + \left(\dfrac{\partial v}{\partial x}\right)^2 + \left(\dfrac{\partial w}{\partial x}\right)^2 \right] \\[3mm] \dfrac{1}{2}\left[\left(\dfrac{\partial u}{\partial y}\right)^2 + \left(\dfrac{\partial v}{\partial y}\right)^2 + \left(\dfrac{\partial w}{\partial y}\right)^2 \right] \\[3mm] \dfrac{1}{2}\left[\left(\dfrac{\partial u}{\partial z}\right)^2 + \left(\dfrac{\partial v}{\partial z}\right)^2 + \left(\dfrac{\partial w}{\partial z}\right)^2 \right] \\[3mm] \dfrac{\partial u}{\partial x}\cdot\dfrac{\partial u}{\partial y} + \dfrac{\partial v}{\partial x}\cdot\dfrac{\partial v}{\partial y} + \dfrac{\partial w}{\partial x}\cdot\dfrac{\partial w}{\partial y} \\[3mm] \dfrac{\partial u}{\partial y}\cdot\dfrac{\partial u}{\partial z} + \dfrac{\partial v}{\partial y}\cdot\dfrac{\partial v}{\partial z} + \dfrac{\partial w}{\partial y}\cdot\dfrac{\partial w}{\partial z} \\[3mm] \dfrac{\partial u}{\partial x}\cdot\dfrac{\partial u}{\partial z} + \dfrac{\partial v}{\partial x}\cdot\dfrac{\partial v}{\partial z} + \dfrac{\partial w}{\partial x}\cdot\dfrac{\partial w}{\partial z} \end{array} \right\},
\tag{18}
$$

where u, v and w are displacements in the x, y and z directions respectively, as before.

The potential energy per unit volume, k, may be written as the sum of the strain energy associated with the small amplitude elastic wave (σ) and the work performed by the initial stress ($\sigma^{(0)}$).

$$
s = \frac{1}{2}\sigma^T \cdot E + \sigma^{(0)T} \cdot E
\tag{19}
$$

If we restrict our analysis to consider only an axial load on the waveguide, which extends in the z direction, we need only retain the initial stress $\sigma_{zz}^{(0)}$. We can assume that the stresses associated with the small amplitude elastic wave are at least an order of magnitude smaller than the axial stress and therefore the product of these stresses and the non-linear strain terms are negligible.

The strain energy can then be written as,

$$
s = \frac{1}{2}[\sigma_{xx}\cdot\varepsilon_{xx} + \sigma_{yy}\cdot\varepsilon_{yy} + \sigma_{zz}\cdot\varepsilon_{zz} + \sigma_{xy}\cdot\varepsilon_{xy} + \sigma_{xz}\cdot\varepsilon_{xz} + \sigma_{yz}\cdot\varepsilon_{yz}] + \sigma_{zz}^{(0)}\cdot E_{zz}.
\tag{20}
$$

All of these terms except the term containing the initial load are already included in the linear strain energy used in the SAFE method. The term containing the initial load can be expanded as follows,

$$
\sigma_{zz}^{(0)}\cdot E_{zz} = \sigma_{zz}^{(0)}\cdot\frac{\partial w}{\partial z} + \sigma_{zz}^{(0)}\cdot\frac{1}{2}[\left(\frac{\partial u}{\partial z}\right)^2 + \left(\frac{\partial v}{\partial z}\right)^2 + \left(\frac{\partial w}{\partial z}\right)^2]
\tag{21}
$$

The term $\sigma_{zz}^{(0)}\cdot\dfrac{\partial w}{\partial z}$ disappears when the variation of the Hamiltonian of the waveguide is taken as in Gavrić (1995) or alternatively when the Lagrange equations are applied as in

Damljanović & Weaver (2004a). This term represents the interaction of the initial stress with the linear strain, which does not feature in the equations of motion for linear systems. Therefore the only term we need to add to the linear strain energy expression, previously used in the SAFE, is,

$$s^{(0)} = \frac{1}{2}\sigma_{zz}^{(0)} \cdot [\left(\frac{\partial u}{\partial z}\right)^2 + \left(\frac{\partial v}{\partial z}\right)^2 + \left(\frac{\partial w}{\partial z}\right)^2].$$

(22)

Substituting the displacement interpolation functions and integrating as before produces the additional strain energy term with the form,

$$k^{(0)} = \frac{1}{2}\sigma_{zz}^{(0)} \cdot \kappa^2 \{-u \quad -v \quad -w\} \cdot \begin{Bmatrix} -u \\ -v \\ -w \end{Bmatrix}.$$

(23)

The form of this term is identical to that of the kinetic energy. Therefore the additional stiffness matrix is proportional to the mass matrix and the equations of motion can be written simply as,

$$M\ddot{u} + \left[\kappa^2 \cdot \left(K_2 + K^{(0)} \right) + \kappa \cdot K_1 + K_0 \right] u = 0 ,$$

(24)

where,

$$K^{(0)} = \frac{\sigma_{zz}^{(0)}}{\rho} M .$$

(25)

As no new matrices have to be created it is trivial to extend existing software to analyse the influence of axial load on the wave propagation characteristics.

4. Computation of dispersion characteristics

If we consider free harmonic vibration (24) and (25) provide the eigenvalue problem, including the initial axial stress, $\sigma_0 = \sigma_{zz}^{(0)}$,

$$\left[\kappa^2 \cdot \left(K_2 + \frac{\sigma_0}{\rho} M \right) + \kappa \cdot K_1 + K_0 \right] u = \omega^2 M u .$$

(26)

To obtain the relationship between wavenumber and frequency it is necessary to specify one of these and solve the eigenvalue problem to obtain the other. If one is interested in the behaviour at a frequency or a range of frequencies these may be computed by complementing (26) with an identity as suggested by Hayashi et al. (2003) and then solving the following complex eigenvalue problem:

$$\begin{bmatrix} K_0 - \omega^2 M & 0 \\ 0 & -\left(K_2 + \dfrac{\sigma_0}{\rho}M\right) \end{bmatrix}\begin{Bmatrix} U \\ \kappa U \end{Bmatrix} + \kappa \begin{bmatrix} K_1 & \left(K_2 + \dfrac{\sigma_0}{\rho}M\right) \\ \left(K_2 + \dfrac{\sigma_0}{\rho}M\right) & 0 \end{bmatrix}\begin{Bmatrix} U \\ \kappa U \end{Bmatrix} = \begin{Bmatrix} 0 \\ 0 \end{Bmatrix}. \tag{27}$$

The wavenumbers that are obtained, by solving this problem, can be real, imaginary or complex and occur in pairs with opposite sign corresponding to waves travelling in opposite directions. If the number of nodes in the model is denoted N, the eigensolution results in $6N$ eigenvalue-eigenvector pairs.

If only the propagating modes are required the real wavenumber may be specified and the eigenvalue problem (26) may be solved. The propagating modes have real frequency, ω and real mode shape ψ. The dispersion characteristics, of the propagating modes, may be obtained by solving this eigenvalue problem for a range of different real wavenumbers and collecting the real frequencies that are produced. This approach is used here. At each specified wavenumber a set of frequency points are obtained but there is no relationship between the frequencies at one wavenumber and those at the next wavenumber. If we want to plot the dispersion curves it is necessary to track the modes from one wavenumber to the next. A technique was developed to track the modes, utilizing the orthogonality property of the mode shapes, as expressed in (28). Equation 28 expresses the mass orthogonality of two arbitrary modes r and s but the stiffness orthogonality could have been used instead.

$$\begin{aligned} \psi_r^{\,T}(\kappa) \cdot M \cdot \psi_s(\kappa) &= 0 \\ \psi_r^{\,T}(\kappa) \cdot M \cdot \psi_r(\kappa) &\neq 0 \end{aligned} \tag{28}$$

It is reasonable to assume that if small wavenumber steps are taken, the mode shapes will not change significantly between steps, and that a mode shape at one wavenumber would almost be mass-orthogonal to those at the next step. The mass-orthogonality of the mode shapes at wavenumber step k, to those at wavenumber step $k+1$, is computed, i.e.,

$$\Theta = \psi^T(\kappa_k) \cdot M \cdot \psi(\kappa_{k+1}). \tag{29}$$

If the wavenumber versus frequency curves have not crossed in this wavenumber interval, the largest terms in the matrix Θ will be the diagonal terms. The presence of an off-diagonal term that is larger than the corresponding diagonal term, indicates that the curves have crossed. For example, if the terms $\Theta_{i,i+1}$ and $\Theta_{i+1,i}$ are larger than the terms $\Theta_{i,i}$ and $\Theta_{i+1,i+1}$ this indicates that the ith mode at wavenumber step k becomes the $i+1$th mode at wavenumber step $k+1$ and that the $i+1$th mode at wavenumber step k becomes the ith mode at wavenumber step $k+1$. This is then taken into account in the numbering of the modes so that a curve for each mode can be plotted. If the complex eigenvalue problem (27) is solved at a range of frequencies a similar approach may be used to plot the dispersion curves of the propagating modes (Loveday, 2008).

The group velocities can be computed using an analytical expression similar to the one presented by Hayashi et al. (2003) for their element formulation. The analytical expression for the group velocity is obtained by differentiating the solution of the eigenvalue problem with respect to frequency,

$$\frac{\partial}{\partial \omega} \psi^T \left[\kappa^2 \left(K_2 + \sigma_0 \frac{M}{\rho} \right) + \kappa K_1 + K_0 - \omega^2 M \right] \psi = 0 , \tag{30}$$

leading to an expression which can be rearranged to obtain the group velocity as the subject.

$$v_g = \frac{\partial \omega}{\partial \kappa} = \frac{1}{2\omega} \frac{\psi^T \left[2\kappa \left(K_2 + \sigma_0 \frac{M}{\rho} \right) + K_1 \right] \psi}{\psi^T M \psi} . \tag{31}$$

The analytical nature of the SAFE method also makes it possible to directly compute the sensitivity of the wavenumber to changes in axial load at a particular frequency as shown in (32).

$$\frac{\partial \kappa}{\partial \sigma} = -\frac{\psi^T \kappa^2 \frac{M}{\rho} \psi}{\psi^T \left[2\kappa \left(K_2 + \sigma_0 \frac{M}{\rho} \right) + K_1 \right] \psi} = -\frac{\kappa^2}{2\omega\rho} \frac{1}{v_g} . \tag{32}$$

In a similar way, the sensitivity of the wavenumber to changes in elastic modulus can be obtained

$$\frac{\partial \kappa}{\partial E} = -\frac{1}{E} \frac{\psi^T \left[\kappa^2 \left(K_2 + \frac{M}{\rho} \sigma_0 \right) + \kappa K_1 + K_0 \right] \psi}{\psi^T \left(2\kappa K_2 + K_1 \right) \psi} = -\frac{\omega}{2E} \frac{1}{v_g} . \tag{33}$$

It is interesting to note the similar form and the role of the group velocity in these sensitivities. This feature was used to compare the change in wavenumber that would be expected to result from a change in temperature via these two mechanisms (Loveday & Wilcox, 2010).

5. Results & discussion

The influence of axial load on the wave propagation in an aluminium rod was computed and compared to an analytical solution. The analytical solution is based on Euler – Bernoulli beam theory and was presented by Chen and Wilcox (2007). The phase velocity v_{ph}, at frequency, ω, for a beam with Young's modulus, E, second moment of area, I, mass per unit length, m, and subjected to a tensile axial load, T, is:

$$v_{ph} = \omega \sqrt{\frac{2EI}{\sqrt{T^2 + 4mEI\omega^2} - T}} . \tag{34}$$

A 1 mm diameter aluminium rod was modelled with SAFE method without axial load and then with a tensile axial load corresponding to 0.1% axial strain. The results obtained are compared to the Euler-Bernoulli beam solution in fig 1. It is clear that the results of the SAFE analysis are practically identical to the Euler-Bernoulli beam model results. This result confirms the accuracy of the formulation and the numerical implementation of the method.

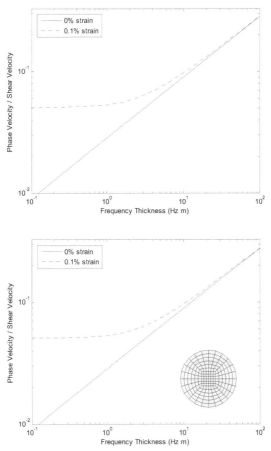

Fig. 1. Influence of strain on 1mm diameter aluminium rod computed by Euler-Bernoulli beam theory (top) and SAFE (bottom).

The ability of the SAFE method to compute the influence of axial loads on complex shapes such as rails was demonstrated by computing the dispersion curves for a UIC60 rail profile. The model used a density of 7700 kg/m3, a Young's modulus of 215 GPa and a Poisson's ratio of 0.3. The axial load applied corresponds to 0.1% strain. This analysis was performed by setting the wavenumber and computing the frequencies of the propagating modes at this wavenumber. The wavenumber was increased in a number of steps and the orthogonality conditions were used to track the evolution of the modes. The wavenumber – frequency curves are shown in fig. 2, which shows numerous propagating modes. The case with no axial load is shown with dashed lines while the solid lines represent the case with axial load. The first four propagating modes, which propagate at all frequencies, are identified in the legend. It is difficult to see the influence of the axial load when the curves are plotted over 50 kHz and 120 rad/m. The second plot shows a zoomed in view and the small influence can be observed in the fundamental horizontal and vertical bending modes in the frequency range between 5 kHz and 10 kHz.

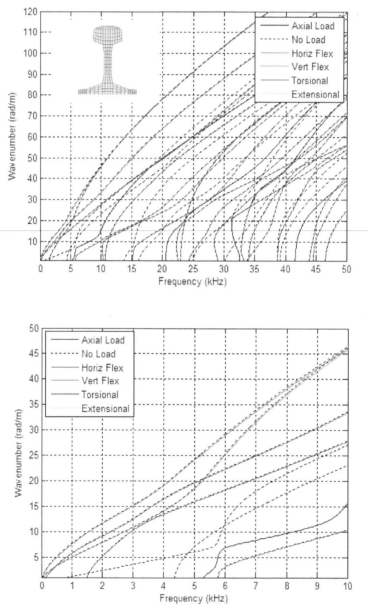

Fig. 2. Influence of strain on dispersion in UIC60 rail over a large frequency range (top) and zoomed in (bottom).

The influence of axial load on the group velocities was computed using (31) and is shown in fig. 3. Again, the influence is small and the zoomed in view is necessary to see the influence.

The group velocity of the two fundamental flexural modes appears to be most sensitive in the frequency range from 3 kHz to 6 kHz. In this range the group velocity changes rapidly with frequency and significant dispersion would be observed in experiments.

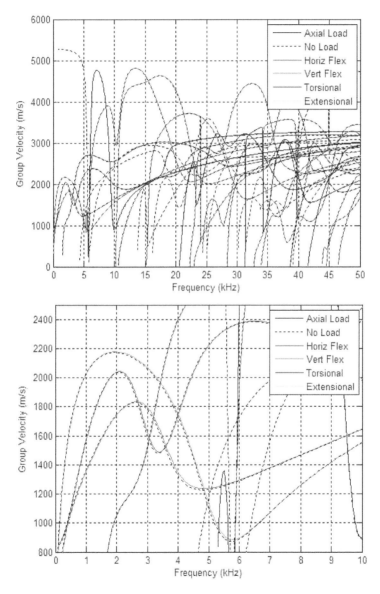

Fig. 3. Influence of strain on group velocity in UIC60 rail over a large frequency range (top) and zoomed in (bottom).

Finally, the sensitivity of the wavenumbers of the propagating modes to the axial load was computed using (32). The sensitivities are plotted in fig. 4 along with the relative sensitivities, which are the sensitivities normalised by the wavenumber.

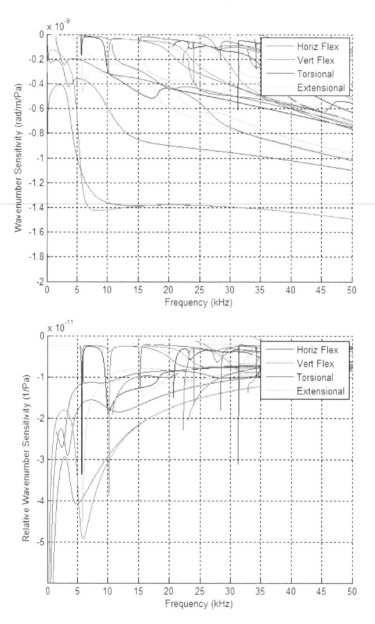

Fig. 4. Sensitivity (top) and relative sensitivity (bottom) of wavenumber to axial load in UIC60 rail.

These sensitivities indicate modes and frequencies where the axial load has an influence on the wavenumber or phase velocity more clearly than the curves in fig. 2. Again, the two fundamental flexural modes are the most sensitive. The relative wavenumber sensitivity provides a measure of how much the wavelength changes relative to the wavelength and if one considers transmitting and then receiving at a number of wavelengths further along the rail one could calculate the amount of phase change that would occur due to the axial load. The relative wavenumber sensitivity is highest at very low frequencies. At slightly higher frequencies the two fundamental flexural modes show maxima in the relative wavenumber sensitivities at approximately 5kHz and 6 kHz. It is interesting to note that these frequencies appear to correspond to minima in the group velocities of these two modes. This information would be useful when trying to design a system to exploit the wavenumber sensitivity to measure the axial load in a rail.

6. Conclusion

The SAFE method has become popular for analysing guided wave propagation in structures with complex geometries. The method, applied to one-dimensional waveguides, was extended to include the presence of an axial load. It was shown that the software modifications required for this extension are trivial. One of the advantages of the SAFE method is that it allows group velocities to be computed analytically. The analytical nature of the method can further be exploited to compute sensitivities analytically and the sensitivity of the wavenumber to axial load was computed in this manner. Results for propagating modes up to 50 kHz in UIC60 rail were computed.

7. Acknowledgment

This work was funded in part by the CSIR SRP Project – TB_2009_19, "Ultrasonic Transducers for Long-Range Rail Monitoring".

8. References

Aalami, B. (1973). Waves in Prismatic Bars of Arbitrary Cross-Section. Journal of Applied Mechanics – Transactions of the ASME, Vol.40, No.4, pp.1067-1077.

Andhavarapu, E.V, Loveday, P.W., Long, C.S. & Heyns, P.S. (2010). Accuracy of Semi-Analytical Finite Elements for Modelling Wave Propagation in Rails. Seventh South African Conference on Computational and Applied Mechanics.

Baronian, V., Lhémery, A. & Bonnet-BenDhia, A-S. (2009). Simulation of non-destructive inspections and acoustic emission measurements involving guided waves. Journal of Physics: Conference Series, Vol.195.

Bartoli, I.; Marzani, A.; Lanza di Scalea, F. & Viola, E. (2006). Modeling Wave Propagation in Damped Waveguides of Arbitrary Cross-Section. Journal of Sound and Vibration, Vol.295, pp. 685-707.

Castaings, M. & Lowe, M. (2008). Finite Element Model for Waves Guided along Solid Systems of Arbitrary Section Coupled to Infinite Solid Media. Journal of the Acoustical Society of America, Vol.123, pp. 696-708.

Chen, F. & Wilcox, P.D. (2006). Load Measurement in Structural Members using Guided Acoustic Waves. Annual Review of Progress in Quantitative NDE, Vol.25, pp. 1461-1468.

Chen, F. & Wilcox, P.D. (2007). The Effect of Load on Guided Wave Propagation. Ultrasonics, Vol.47, pp. 111-122.

Damljanović, V. & Weaver, R.L. (2004a). Propagating and Evanescent Elastic Waves in Cylindrical Waveguides of Arbitrary Cross Section. Journal of the Acoustical Society of America, Vol.115, No.4, pp. 1572–1581.

Damljanović, V. & Weaver, R.L. (2004b). Forced Response of a Cylindrical waveguide with Simulation of the Wavenumber Extraction Problem. Journal of the Acoustical Society of America, Vol.115, No.4, pp. 1582-1591.

Damljanović, V. & Weaver, R.L. (2005). Laser Vibrometry Technique for Measurement of Contained Stress in Railroad Rail. Journal of Sound and Vibration, Vol.282, pp. 341–366.

Gavrić, L. (1995). Computation of Propagative Waves in Free Rail using a Finite Element Technique. Journal of Sound & Vibration, Vol.185, No.3, pp. 531–543.

Gazis, D.C. (1959). Three-Dimensional Investigation of the Propagation of Waves in Hollow Circular Cylinders. Journal of the Acoustical Society of America, Vol.31, pp. 568-578.

Hayashi, T.; Song, W-J. & Rose, J.L. (2003). Guided Wave Dispersion Curves for a Bar with an Arbitrary Cross-Section, a Rod and Rail Example. Ultrasonics, Vol.41, pp.175-183.

Lagasse, P.E. (1973). Finite Element Analysis of Piezoelectric Elastic Waveguides. IEEE Transactions on Sonics and Ultrasonics, Vol.SU-20, pp. 354-359.

Loveday, P.W. (2000). Development of Piezoelectric Transducers for a Railway Integrity Monitoring System, Smart Structures and Materials 2000: Smart Systems for Bridges, Structures, and Highways, Proceedings of SPIE Vol. 3988, pp.330-338.

Loveday, P.W. (2008). Simulation of Piezoelectric Excitation of Guided Waves Using Waveguide Finite Elements. IEEE Transactions on Ultrasonics, Ferroelectrics and Frequency Control, Vol.55, No.9, pp.2038-2045.

Loveday, P.W. (2009). Semi-Analytical Finite Element Analysis of Elastic Waveguides Subjected to Axial Loads. Ultrasonics, Vol.49, pp. 298-300.

Loveday, P.W. & Wilcox, P.D. (2010). Guided Wave Propagation as a Measure of Axial Loads in Rails. Health Monitoring of Structural and Biological Systems, Proceedings of the SPIE, Vol.7650.

Lowe, M.J.S.; Alleyne, D.N. & Cawley, P. (1998). Defect Detection in Pipes Using Guided Waves. Ultrasonics, Vol.36, pp. 147-154

Nilsson, C.M. & Finnveden, S. (2007). Input power to waveguides calculated by a finite element method. Journal of Sound and Vibration, Vol.305, pp. 641–658.

Predoi Mihai, V., Castaing, M. & Hosten, B. (2007). Wave Propagation along Transversely Periodic Structures. Journal of the Acoustical Society of America, Vol.121, pp. 1935-1944.

Rose, J. L. (1999). Ultrasonic Waves in Solid Media, Cambridge University Press.

Rose, J.L. (2002). Standing on the Shoulders of Giants: An Example of Guided Wave Inspection. Materials Evaluation, pp. 53-59.

Ryue, J., Thompson, D.J., White, P.R. & Thompson, D.R. (2008). Investigations of propagating wave types in railway tracks at high frequencies. Journal of Sound and Vibration, Vol.315, pp. 157-175.

Thompson, D. J (1997). Experimental analysis of wave propagation in railway tracks. Journal of Sound and Vibration, Vol.203, pp. 867–888.

10

Investigation of Broken Rotor Bar Faults in Three-Phase Squirrel-Cage Induction Motors

Ying Xie

Harbin University of Science and Technology,
China

1. Introduction

It is well understood that squirrel-cage induction motors are rugged, reliable, cheap, and thus widely used in industrial and manufacturing processes. However, electrical and mechanical faults pose a particular challenge to the industry and end users which often interrupts the productivity and requires maintenance. In literature, rotor faults have been shown to account for a large portion of induction motor failures, sometimes they are the single biggest cause of failure in the field. Rotor bar faults generally arise from repeated operating stresses which can be electrical, mechanical, thermal or environmental by nature. The causes of rotor bar and end-ring breakage include: (a) magnetic stresses caused by electromagnetic forces, (b) thermal stresses due to abnormal operating duty, including overload and unbalance, (c) inadequate casting, fabrication procedures or overloading, (d) contamination and abrasion of rotor because of poor operating conditions, (e) lack of maintenance. Most failures will increase the current and stress in the adjacent bars, progressively deteriorating the rotor part and degrading the motor's overall performance. Without doubt, it is of great importance to appreciate the mechanisms and characteristic changes of broken bar faults. Furthermore, an online fault diagnostic system is highly desired to meet the reliability requirements, as well as reducing the costs of maintenance and increasing field efficiency.

2. Broken bar faults in squirrel-cage induction motors

During the past twenty years, there have been continuing efforts at studying and diagnosing faults in induction motors and, in particular, substantial research work is devoted to induction motor bar breakages and the development of non-intrusive diagnostic techniques (Elkasabgy et al., 1992; Bellini et al., 2001; Said et al., 2000). Some research work was based on the finite element (FE) techniques (Mohammed et al., 2006; Mirafzal & Demerdash, 2004; ying xie, 2009; Sprooten & Maun, 2009; Bentounsi & Nicolas, 1988; ying xie, 2010), and more information may be retrieved for diagnostic purpose. It is well established to use line currents as an indicative parameter (Kliman et al., 1988) which can provide insight into the basis of a non-invasive condition monitoring system for the early detection. Other research effort has been focused on the motor current signature analysis (Costa, et al., 2004; Walliser & Landy, 1994; Bacha et al., 2004; Thomson & Fenger, 2001) in order to detect electrical and mechanical faults in induction motors. Another issue reported in literature is the temperature-rise-related failure which has also received much research attention. For example, some research has been conducted on totally enclosed fan-cooled (TEFC) induction motors by thermal sensitivity analysis (Mueller et al., 1995; Boglietti et al., 2005; Staton et al., 2005;). In this reference, the thermal design issues were reviewed and optimisation design algorithms were also developed. References (Alberti & Bianchi, 2008) propose a coupled thermal–magnetic analysis of an induction motor with the primary goal of achieving a rapid and accurate prediction of the IM performance. The heating problem of a motor when one of rotor bars is totally broken was simulated in some papers (Casimir et al., 2004); and some papers investigated the heating characteristics and heat distribution of the motor with healthy and broken rotors (Cho et al., 1992; Lopez-Fdez et al., 1999; Antal & Zawilak, 2005). Indeed, there are some technical challenges when analysing electrical motor thermal fields under broken-bar fault conditions although many papers have covered this area of research. In particular, there is little work on the influence of breaking bars on the temperature-rise of electrical motors using quantitative methods. This chapter will bridge the gap. This chapter is also set out to discuss early diagnostic techniques to identify the faults.

3. Experimental setup

A dedicated experimental test bench has been designed for testing squirrel-cage induction motor faults. One stator and three originally identical rotors have been employed to study the behaviours of the induction motor with or without broken bars. Prior to the start of the testing process, two out of the three rotors are damaged deliberately by drilling holes in the bars on all their depth and used with the same stator to ensure the testing accuracy. The healthy rotor is considered here as a reference. The laboratory test setup and broken bar rotors are shown in Fig. 1.

Fig. 1. The experimental setup and the three rotors.

4. Search coil and voltage detection techniques

The measured value of air-gap flux density can be obtained by the search coil technology, and it can be applied to motors operating under different loads. The search coil is inserted around the stator tooth tip, as is shown in Fig. 2. The value of air-gap flux density is obtained by analyzing the induced voltage waveform in the coil, which could help to detect the presence of broken bars.

The technique uses a search coil mounted on the internal stator tooth tip and the analysis of the induced voltage waveform to detect the presence of broken bars. The induced voltage in search coil is given by

$$E_i = 2B_iLv = 2\frac{B_{im}}{\sqrt{2}}L\frac{\pi Dn_s}{60} \tag{1}$$

where B_{im} is the amplitude of the fundamental component and the i th harmonic component, E_i is the effective value of voltage induced by the fundamental air-gap flux and the i th harmonic air-gap flux, D is the inner diameter of the stator core, n_s is the synchronous speed, L is the effective length of the search coil.

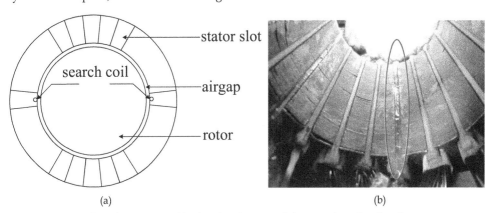

(a)   (b)

Fig. 2. The search coil positions. (a) The sketch map of the search coil technology, (b) measuring position in the experiment

Under rated load conditions, the air-gap field is dissymmetrical in the case of broken bars and the harmonic components of air-gap flux density vary significantly. The third harmonic components for the case of a faulty motor are pronouncedly higher and ripple more severely than that of the healthy condition. The appearance of a faulty third harmonic component is clearly an effective method of detecting broken bars, and the test results showed the faulty third harmonic component of air-gap flux density becomes more significant as the number of the broken bars is increased. The results of the first and third harmonic of the air-gap flux density are shown in Table I for the sample simulation and experiment. (The first harmonics for the case of normal and abnormal conditions from experiment are also fluctuated with time, and the maximal values are listed in the Table 1.) (Weili Li et al., 2007)

	1th harmonic (t=1s)		3th harmonic (t=1s)	
	Experimental	Simulated	Experimental value range	Simulated
Healthy bars	0.6534 (T)	0.6414 (T)	0.0008-0.0168(T)	0.0113 (T)
A broken bar	0.6601 (T)	0.6437 (T)	0.0034-0.0386(T)	0.0216 (T)
Two adjacent broken bars	0.6887 (T)	0.6480 (T)	0.0042-0.0848(T)	0.0391 (T)

Table 1. The fundamental and third harmonic component of air gap flux density comparison for simulation and experiment at rated load

5. Influence of broken bar faults on the magnetic field distribution

The following assumptions have been made in the solution procedure.

- Displacement current is neglected because the frequency of the source is very low.
- The rotor bars are insulated from the rotor core, and there is no direct electrical contact between the rotor bars and the rotor core.
- The leakage on the outer surface of the stator and the inner surface of the rotor is neglected.
- The 2-D domain is considered, and the magnetic vector potential and the current density have only the axial z component.

The 2-D model of the motor is employed, and the fundamental equation describing the space and time variations of the vector potential has the following form.

$$\begin{cases} D: \dfrac{\partial}{\partial x}\left(\dfrac{1}{\mu}\dfrac{\partial A_z}{\partial x}\right) + \dfrac{\partial}{\partial y}\left(\dfrac{1}{\mu}\dfrac{\partial A_z}{\partial y}\right) = -J_z + \sigma\dfrac{\partial A_z}{\partial t} \\ \Gamma_1 : A_z = 0 \end{cases} \tag{2}$$

where D is the region of analysis, Γ_1 is the outside circumferential of the stator and inside circumferential of the rotor (the Dirichlet boundary conditions), J_z is current density, σ is the conductivity of the conductors, A_z is magnetic vector potential, μ is the permeability of the material (Ning Yuquan,2002; Tang Yunqiu,1998; Yan Dengjun et al., 2003; Bangura & Demerdash 1999; Gao Jingde et al., 1993; ying xie, 2009).

The distribution of magnetic field of the motor for the case of no broken bars is symmetrical under the rated load conditions and the locked rotor conditions, while the symmetry of magnetic field distribution is distorted in the case of broken bars. Samples of flux density distributions across the cross-section of the motor are shown in Fig. 3-4. From them, we know that the magnetic saturation around the broken bars is more severe than that of the rated load. In addition to changes in the broken bar regions, the field distributions at other positions in the stator and rotor core are also distorted and increased to some extent, while these are less significant in the case of the rated load.

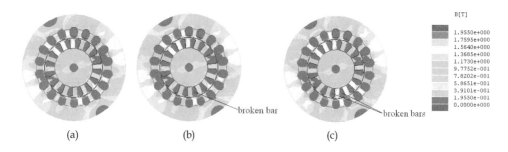

Fig. 3. The flux density distribution at rated load. (a) healthy rotor, (b) one broken bar, (c) two bar broken

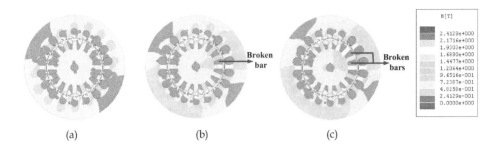

(a)   (b)   (c)

Fig. 4. The flux density distribution at standstill. (a) healthy rotor, (b) one broken bar fault, (c) two bar broken fault

In order to study the higher degree of magnetic saturation around broken bars, the magnetic flux density waveforms of A, B and C points are given in this paper. The three different positions which are adjacent to broken bars in the stator and rotor core are shown in Fig.5.

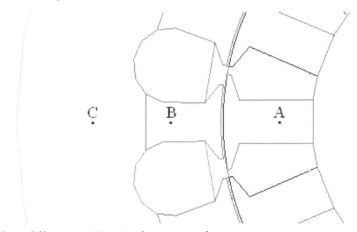

Fig. 5. The three different positions in the stator and rotor core.

The typical time variation flux density waveforms of x and y component at positions A. B and C are shown in Fig.6. One can notice that the flux density change is periodic and calm on the condition of healthy rotor, however, the flux density fluctuations with time after broken bar fault. A comparison of flux density plot between the healthy cage and the broken bars fault demonstrates the harmonic components of the flux density on the different positions for broken bars are greater than those for healthy rotor. To specify this statement, the magnetic flux frequency spectrum at selected position B is analyzed by harmonic analysis procedure, see Fig.7. From it we can know that the flux density contains higher spatial harmonics in the case of broken bars.

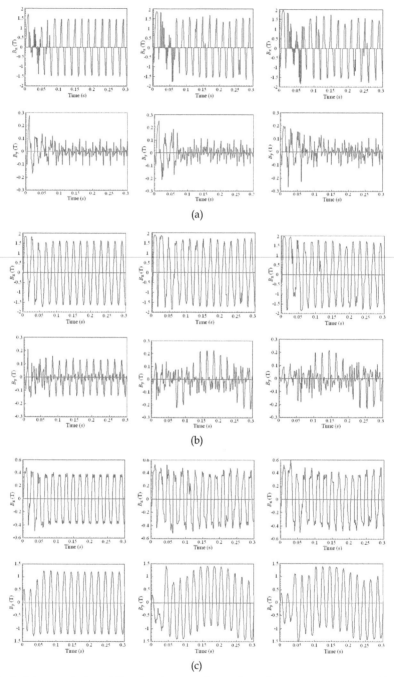

Fig. 6. Typical time variation of flux density waveform of x and y component at rated load with and without broken bar fault. (a) positions A, (b) positions B , (c) positions C (ying xie,2009)

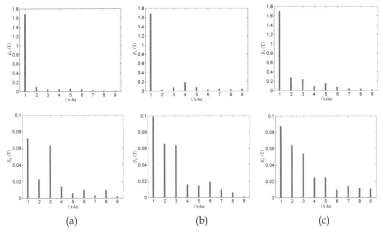

Fig. 7. The flux density frequency spectrum at position B at rated load. (a) healthy motor cage, (b) a one-broken bar fault, (c) a continuous two-broken-bar

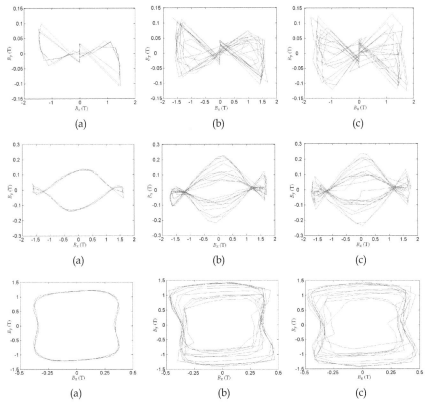

Fig. 8. Elliptical flux density vector waveform at position A, B and C at rated load. (a) healthy motor cage, (b) a one-broken bar fault, (c) a continuous two-broken-bar

Fig. 8 is the elliptical flux density at positions A, B and C, from it, we can see that the trace of elliptical flux density in case of healthy motor is nearly the same; however it is disorderly and unsystematic when broken bar fault happened. All radii of elliptical flux density vector for broken bars are greater than those for the healthy case, which is due to the local heavy magnetic saturation appearing in the vicinity of the bar breakages.

6. Operation characteristics of induction motors with broken bar fault

The effect of the broken bar in three-phase cage-rotor induction motors on the motor's operating performances is investigated under both the rated load conditions and the locked rotor conditions. A 2-D Time-Stepping Coupled Finite Element Method (TSCFEM) is employed for predictive characterization of rotor broken bars in induction motors. Simulation results based on detailed theoretical analysis are confirmed by the experimental results.

6.1 Stator currents

In the generalized rotating field theory, a backward-rotating field can be produced by the broken rotor bar faults and then lower sideband components in the stator current spectrum at double slip-frequency is introduced. Figs. 9-10 show experimental and simulated transient phase currents at rated load. One can notice that the amplitude of stator current fluctuations with time compared to that in the healthy cage. However, while the tests are performed at standstill, the fault-specific sideband components of stator currents do not appear near the fundamental component. Therefore, the stator current for healthy rotor at standstill is very similar to that for faulty rotors. Figs. 11-12 show experimental and simulated stator currents at standstill with healthy and faulty rotors for comparison.

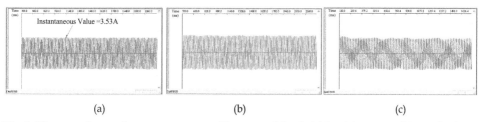

(a)   (b)   (c)

Fig. 9. The experimental stator current profile at rated load. (a) healthy rotor, (b) one broken bar, (c) two broken bars

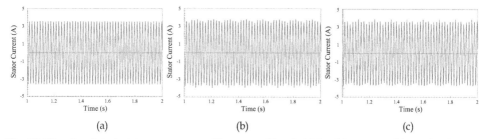

(a)   (b)   (c)

Fig. 10. The simulated stator current profile at rated load. (a) healthy rotor, (b) one broken bar, (c) two broken bars (ying xie 2009)

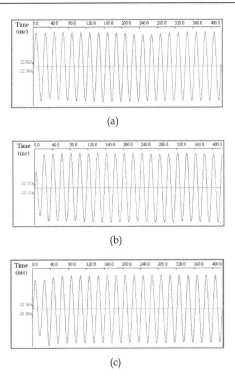

(c)

Fig. 11. The experimental stator current waveforms at standstill. (a) healthy rotor, (b) one broken bar, (c) two broken bars

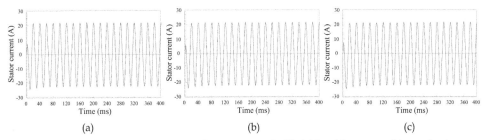

Fig. 12. Simulated stator current waveforms at standstill. (a) healthy rotor, (b) one broken bar, (c) two broken bars

6.2 Rotor-bar currents

When the rotor is rotating, each rotor bar passes every stator slot, so that each bar will be equally influenced by all the stator-driven flux waves, and all the currents of the rotor bars at rated load are sensibly uniform around the rotor periphery. Fig. 13 shows rotor-bar currents at rated load. It can be seen that the amplitude of the adjacent bars has the highest value in the bars next to the broken ones, this explains why and how bar damage propagates. The currents in bars far away from the broken bars remain almost the same.

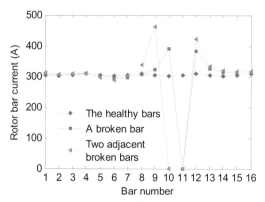

Fig. 13. The simulated rotor bar current at rated load.

At standstill, however, the current amplitude of the trouble-free rotor varies with position around the rotor periphery and is not equal, which is different from the rated load. The variation of the rotor current in fault at standstill in accordance with it at rated load and the current amplitude at standstill increases more serious, which can be seen from Figure 14.

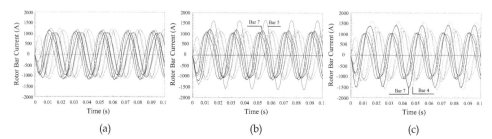

Fig. 14. The simulated rotor current at standstill: (a) healthy cage, (b) one broken bar, (c) two broken bars

6.3 Magnetic force on the rotor

There have been a variety of methods for calculating local magnetic forces, i.e. the methods based on the virtual work principle, on the Maxwell stresses or on the forces acting on equivalent sources (magnetizing current or magnetic charges). In this section, the method of virtual work is employed in the process of the magnetic force calculation. There is the magnetic force on the rotor bars because of the induced current in the bars. In the two-dimensional magnetic field, the magnetic force can be expressed as follows.

$$f_k = \iint_{S_k} Jl \times B_k \cdot dS \tag{3}$$

where k is unit number, f_k is the magnetic force of the unit k, S_k is the area of the unit k, J is the induced current density on the rotor bars, l is the length of the bars, B_k is the magnetic flux density of the unit k. The magnetic force corresponding to Eq.3 may be simplified, and it becomes

$$\begin{cases} f_{t,k} = B_{n,k}S_kJl \\ f_{n,k} = B_{t,k}S_kJl \end{cases} \tag{4}$$

where $f_{t,k}$, $f_{n,k}$ are the tangential component and the normal component of magnetic force, $B_{t,k}$ and $B_{n,k}$ are the tangential component and the normal component of magnetic flux density respectively. Therefore, the magnetic force of the unit k is

$$f_k = if_{t,k} + jf_{n,k} \tag{5}$$

In this section, the magnetic force distribution on the rotor bar at rated load and at standstill is computed, and the position of broken bars is shown in the Fig.15. The magnetic force distributions on the rotor bar at rated load and at standstill are computed by the FE method and the results are shown in the Fig. 16-17 for comparison. It can be noticed that the bars with the highest magnetic force are those immediately adjacent to the broken bars, whether the motor operating under rated load conditions or standstill conditions. Consequently, such non-uniform distribution of the force inevitably leads to excessive mechanical stress in the bars, and the bars would become more susceptible to additional wearing and eventual breaking.

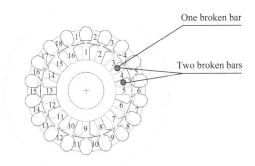

Fig. 15. The serial number of the stator tooth and rotor tooth.

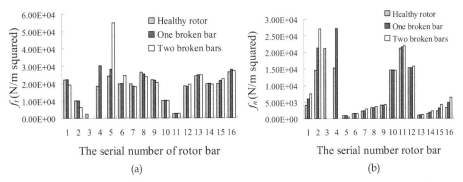

Fig. 16. The magnetic force distribution of every rotor bar at rated load. (a)Tangential component, (b) Normal component

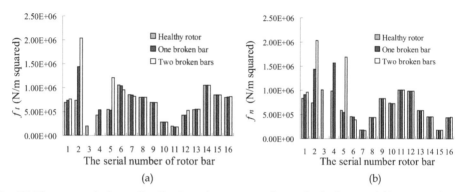

Fig. 17. The magnetic force distribution of every rotor bar at locked rotor. (a)Tangential component, (b) Normal component

6.4 Torques

The torque variation at rated load is given in Fig.18, and the torque is smooth at no fault, and torque ripple can be observed in faulty conditions. The torque tendency at rotor-locked conditions is different to that at rated load condition, see Fig. 19. The torque waveforms are almost identical. Through further observations, the average torque is reduced at locked-rotor conditions (from 12.28, 11.23 to 10.22 Nm, respectively). It becomes clear that the average torque continues to decrease, impacting on the loading capability of the motor.

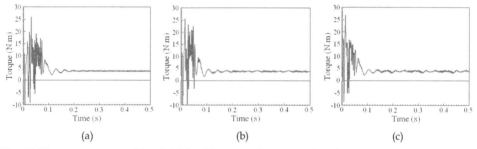

Fig. 18. The torque at rated load. (a) healthy cage, (b) one broken bar, (c) two broken bars (ying xie, 2009)

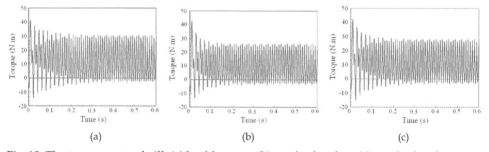

Fig. 19. The torque at standstill: (a) healthy cage, (b) one broken bar, (c) two broken bars

6.5 Core loss of the motor

The variation of iron core loss at rated load with time is shown in Fig.20. For motor with healthy bars, the core loss of stator is stable under steady state. When broken bars fault happened, the starting core loss of stator is significantly higher than normal motor, and the core loss is fluctuant with time rather than smooth under steady state. The amplification of this distortion is directly related to the number of broken bars, and this was mainly due to deformation of electromagnetic field deduced by broken bars fault, and the magnetic saturation and higher harmonic component around the broken bars.

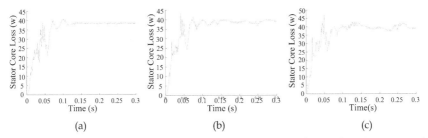

| (a) | (b) | (c) |

Fig. 20. Variations of stator core losses versus time before and after broken bars at rated load. (a) healthy motor cage, (b) a one-broken-bar fault, (c) a continuous two-broken- bar (ying xie, 2009)

On figure 21-22, we present the variation of stator and rotor iron core loss with time at locked-rotor for healthy and broken bars faulty condition respectively.

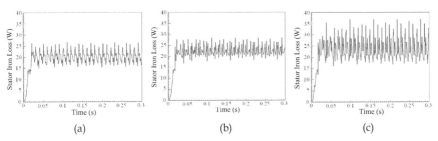

| (a) | (b) | (c) |

Fig. 21. Typical time variation of stator core loss at locked rotor. (a) healthy motor cage, (b) a one-broken-bar fault, (c) a continuous two-broken- bars

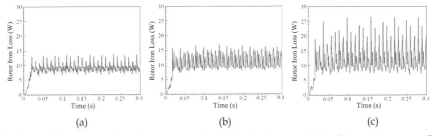

| (a) | (b) | (c) |

Fig. 22. Typical time variation of rotor core loss at locked rotor. (a) healthy motor cage, (b) a one-broken-bar fault, (c) a continuous two-broken- bars

From it we can note the stator and rotor core losses are fluctuant with time whatever the motor is normal or not, which is different from the rated load conditions, for motor with healthy bars, the core loss of stator and rotor is stable under steady state when the motor is operating in the rated load. When broken bars fault happens, the core losses of stator and rotor are significantly higher than normal motor at standstill, and the fluctuation is more intense. In addition, the rotor core losses can not be ignored at standstill.

7. Influence of broken bar faults on the thermal field distribution

For TEFC (Totally Enclosed Fan-Cooled) induction motor, the 2-D thermal analysis is well accepted. Then the difficulty in calculating the thermal field is reduced to some extent and the simulation time is beneficially reduced. In terms of the calculation results of electromagnetic field and some empirical formulas, the heat losses can be obtained. The steady temperature distributions of the motor operating at the rated load are calculated shown as Fig.23.

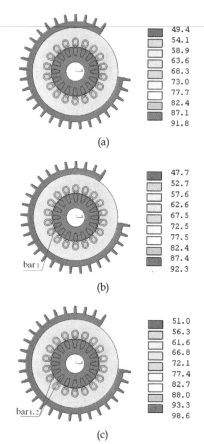

Fig. 23. Temperature distribution of solving region: (a) healthy motor cage; (b) Bar 1 broken; (c) Bar 1 and Bar 2 broken (ying xie 2010)

It can be seen that the rotor temperature is highest, and the temperature distribution tendencies of the faulty conditions are similar to that of the motor with healthy rotor. Therefore the broken bar fault has an unobvious influence on the total temperature distribution tendency of the motor.

Fig.24 are the steady rotor temperature distributions of the motor at the above three states. The rotor temperature distribution of the motor with a healthy rotor is not complete symmetry because of the quasi-stationary-state treatment of the air-gap and the incomplete symmetry of the motor house. But the whole rotor solving region is quite small which is due to the large thermal conductivities of the rotor core and rotor bar. It can be found that the lowest temperatures are in the positions of broken bars in the whole rotor solving region from Fig.24 (b) and (c). It indicates that with the increase of the broken bar number, the temperature-rise at the same position of the motor increases. It can be predicted that the temperature-rises of the stator windings and the rotor will increase dramatically in the case of the motor with serious adjacent broken bars fault.

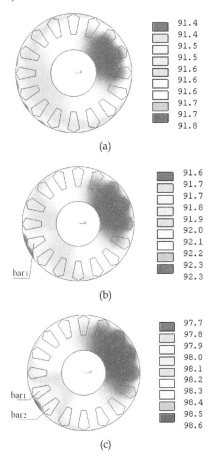

Fig. 24. Rotor temperature distribution. (a) healthy motor cage, (b) a bar 1 broken, (c) bar 1 and bar 2 broken (ying xie 2010)

The air-gap temperature distribution along radial is given in Fig. 25. From it the temperature gradient of the air-gap along radial is rather large. The temperature distribution throughout stator slot along radial of the motor cross section is given as Fig. 26.

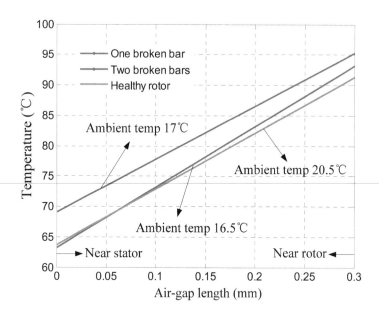

Fig. 25. The air-gap temperature distribution along radial

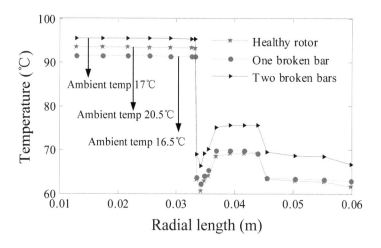

Fig. 26. Temperature distribution throughout radial

8. Conclusions

In this chapter, the application of a Time-Stepping Coupled Finite Element Method for predictive characterization of effects of rotor broken bars has been presented in a comprehensive fashion. The FE analysis has clearly showed that the effect of the broken-bar fault on motor electromagnetic, mechanical performance, and temperature field. Core losses and current profiles of the stator and rotor, the magnetic force and torque in the rotor bar are also affected by the presence of broken bar faults and the motor performance would deteriorate as the number of broken rotor bars increases. Simulation results based on detailed theoretical analysis are validated by the experimental results.

Experimental test and simulation results have illustrated the reason why the broken bar faults are severe and the likelihood of fault propagation to the adjacent bars. From the results in the work, one can appreciate that the broken bar position has a great impact on the motor's operation, especially on the stator current and starting torque. This further confirms the capability of the proposed numerical models which have accounted for the impact of harmonic components of air-gap flux density. Clearly, this research has also highlighted a necessity for advanced online diagnostic techniques to detect the broken bar faults since these are a common and severe type of mechanical faults to break down the induction motors in service.

However, it needs to point out that this chapter has taken use of a 2-D finite element method to analyze the induction motor's electro-magnetic, thermal, mechanical performance, which is proved to be suitable. If more complex problems are involved such as overhang region bar faults, a 3-D finite element method may be required. This is the further work of this research.

9. Acknowledgement

This work was supported in part by National Natural Science Foundation of China (51107022), Specialized Research Fund for the Doctoral Program of Higher Education (20102303120001), and China's Postdoctoral Science Foundation (20100480891).

10. References

Alberti, L. & Bianchi, N. (2008). a Coupled Thermal-Electromagnetic Analysisfor a Rapid and Accurate Prediction of IM Performance. *IEEE Transactions on Industrial Electronics*, Vol. 55, No. 10, (October 2008), pp. 3575–3582, ISSN 0278-0046

Antal, M. & Zawilak, J. (2005). Coupling Magneto-thermal Field of Induction Motor with Broken Rotor Bars. *Maszyny Elektryczne*, Vol.72, (2005), pp. 267-272, ISBN 83-204-0335-9

Bacha, K.; Gossa, M., Capolino, G.-A. (2004). Diagnosis of Induction Motor Rotor Broken Bars. *2004 IEEE International Conference on Industrial Technology*, pp.979-984, ISBN 0-7803-8662-0, Hammamet, Tunisia, December 8-10,2004

Bangura, J.F. & Demerdash, N.A. (1999). Diagnosis and Characterization of Effects of Broken Bars and Connectors in Squirrel-cage Induction Motor by Time-stepping Coupled

FE State Space Modeling Approach. *IEEE Trans. EC.*, Vol.14, (April 1999), pp.1167-1175, ISSN 0885-8969

Bellini, A.; Filippetti, F., Franceschini, G., Tassoni, C., Kliman, G.B. (2001). Quantitative Evaluation of Induction Motor Broken Bars by Means of Electrical Signature Analysis. *IEEE Trans on Industry Applications*, Vol.37, No.5, (2001), pp. 1248-1255, ISSN 0093-9994

Bentounsi, A. & Nicolas, A. (1988). on Line Diagnosis of Defaults on Squirrel Cage Motor Using FEM. *IEEE Trans. Magnetics*, Vol.34, No.5, (September 1998),Part:1, pp. 3511-3514, ISSN 0018-9464

Boglietti, A.; Cavagnino, A., Staton, D.A. (2005). TEFC Induction Motors Thermal Models: A Parameter Sensitivity Analysis. *IEEE Trans on Industry Applications*, Vol.41, No.3, (May/June 2005), pp. 756-763, ISSN 0093-9994

Casimir, R.; Bouteleux, E., Yahoui, H., Clerc, G., Henao, H., Delmotte, C., Capolino, G.-A., Rostaing, G., Rognon, J.-P., Foulon, E., Loron, L., Razik, H., Didier, G., Houdouin, G., Barakat, G., Dakyo, B., Bachir, S., Tnani, S., Champenois, G., Trigeassou, J.-C., Devanneaux, V., Dagues, B., Faucher, J.(2004). Comparison of Modeling Methods and of Diagnostic of Asynchronous Motor in Case of Defects. *International Power Electronics Congress - CIEP, 9th IEEE International Power Electronics Congress - Tehcnical Proceedingss*, pp. 101-108, ISBN 0-7803-8790-2, Celaya, Mexico, October, 2004

Cho, K.R.; Lang, J.H., Umans, S.D. (1992). Detection of Broken Rotor Bars in Induction Motors Using State and Parameter Estimation. *IEEE Transactions on Industry Applications*, Vol. 28,No.3, (May/Jun 1992), pp. 702-709, ISSN 0093-9994

Costa, F.F.; de Almeida, L.A.L., Naidu, S.R., Braga-Filho, E.R., Alves, R.N.C. (2004). Improving the Signal Data Acquisition in Condition Monitoring of Electrical Machines. *IEEE Trans on Instrumentation and Measurement*, Vol. 53, (August 2004), pp.1015-1019, ISSN 0018-9456

Elkasabgy, N.M.; Eastham, A.R., Dawson, G.E. (1992). Detection of Broken Bars in the Cage Rotor on an Induction Machine. *IEEE Trans on Industry Applications*, Vol.28, No.1, (1992), pp. 165 –171, ISSN 0093-9994

Gao Jingde; Wang Xiangheng, Li Fahai. (1993). *Analysis of AC Machines and Their Systems*, Tsinghua University Press, ISBN 7-302-01251-2, Beijing

Kliman, G.B.; Koegl, R.A., Stein, J., Endicott, R.D., Madden, M.W. (1988). Noninvasive Detection of Broken Rotor Bars in Operating Induction Motors. *IEEE Trans on Energy Conversion*, Vol.3, No.4, (December 1988), pp. 873-879, ISSN 0885-8969

Lopez-Fdez, X.M.; Donsion, M.P., Cabanas, M.F., Melero, M.G., Rojas, C.H. (1999). Thermal performance of a 3-phase induction motor with a broken bar, *SDEMPED'99 Record*, pp. 529-533, ISBN 978-0-7803-9124-6, Gijón, Spain, September 1999

Mirafzal, B. & Demerdash, N.A.O. (2004). Induction Machine Broken-bar Fault Diagnosis Using the Rotor Magnetic Field Space-vector Orientation. *IEEE Trans. Industry Applications*, Vol.40, No.2, (February 2004), pp. 534–542, ISSN 0093-9994

Mohammed, O.A.; Abed, N.Y., Ganu, S. (2006). Modeling and Characterization of Induction Motor Internal Faults Using Finite-Element and Discrete Wavelet Transforms. *IEEE Trans. Magnetics*, Vol.42, No.10, (October 2006), pp. 3434–3436, ISSN 0018-9464

Mueller, M.A.; Williamson, S., Flack, T.J., Atallah, K., Baholo, B., Howe, D., Mellor, P.H. (1995).Calculation of Iron Losses from Time-stepped Finite-element Model of Cage Induction Machines. *IEEE Conference Publication*, No.412, (September 1995), pp. 88-92, ISSN 0537-9989

Ning Yuquan. (2002). Faults Detection and On-line Diagnosis Calculating Parameter in Squirrel Cage Induction Motors with Broken Bars and End Ring Connections. *Proceedings of the Chinese Society for Electrical Engineering*, Vol.2, No.10, (May 2002), pp. 97-103, ISSN 0258-8013

Said, M.S.N.; Benbouzid, M.E.H., Benchaib,A. (2000). Detection of Broken Bars in Induction Motors Using an Extended Kalman Filter for Rotor Resistance Sensorless Estimation. *IEEE Trans on Energy Conversion*, Vol.15, No.1, (March 2000), pp. 66-70, ISSN 0885-8969

Sprooten, J. & Maun, J.-C.(2009). Influence of Saturation Level on the Effect of Broken Bars in Induction Motors Using Fundamental Electromagnetic Laws and Finite Element Simulations. *IEEE Trans. Energy Conversion*, Vol.24, No.3, (September 2009), pp. 557–564, ISSN 0885-8969

Staton, D.; Boglietti, A., Cavagnino, A. (2005). Solving the More Difficult Aspects of Electric Motor Thermal Analysis in Small and Medium Size Industrial Induction Motors. *IEEE Trans on Energy Conversion*, Vol.20, No.3, (September 2005), pp.620-628, ISSN 0885-8969

Tang Yunqiu. (1998). *Electromagnetic Field in Electric Machine*. Science Press, ISBN 7-03-005296-X, Beijing

Walliser, R.F. & Landy, C.F. (1994). Determination of Interbar Current Effects in the Detection of Broken Rotor Bars in Squirrel Cage Induction Motor. *IEEE Trans on Energy Conversion*, Vol.9, No.1, (March 1994), pp. 152-158, ISSN 0885-8969

Weili, Li; Xie Ying, Shen Jiafeng, Luo Yingli. (2007). Finite-Element Analysis of Field Distribution and Characteristic Performance of Squirrel-Cage Induction Motor With Broken Bars.IEEE Transactions on Magnetics, Vol.43, No.4, (April 2007), pp. 1537-1540, ISSN 0018-9464

W.Thomson. & M.Fenger. (2001). Current Signature Analysis to Detect Induction Motor Faults. *IEEE Industry Applications Magazine*, Vol.7, No.4, (July/August 2001), pp. 26-34, ISSN 0093-9994

Xie Ying. (2009). Characteristic Performance Analysis of Squirrel Cage Induction Motor with Broken Bars. *IEEE Trans. Magnetics*, Vol.45, No.2, (February 2004), Part:1, pp. 759–766, ISSN 0018-9464

Xie Ying. (2010). Performance Evaluation and Thermal Fields Analysis of Induction Motor with Broken Rotor Bars Located at Different Relative Positions. *IEEE Trans. Magnetics*, Vol.46, No.5, (May 2010), pp. 1243–1250, ISSN 0018-9464

Yan Dengjun; Liu Ruifang, Hu Mingqiang, Li Xunming. (2003). Transient Starting
     Performance of Squirrel Cage Induction Motor with Time-stepping FEM. *Electric
     machines and control*, Vol. 7,(July 2003), pp. 177-181, ISSN 1007-449X

Permissions

The contributors of this book come from diverse backgrounds, making this book a truly international effort. This book will bring forth new frontiers with its revolutionizing research information and detailed analysis of the nascent developments around the world.

We would like to thank David Moratal, for lending his expertise to make the book truly unique. He has played a crucial role in the development of this book. Without his invaluable contribution this book wouldn't have been possible. He has made vital efforts to compile up to date information on the varied aspects of this subject to make this book a valuable addition to the collection of many professionals and students.

This book was conceptualized with the vision of imparting up-to-date information and advanced data in this field. To ensure the same, a matchless editorial board was set up. Every individual on the board went through rigorous rounds of assessment to prove their worth. After which they invested a large part of their time researching and compiling the most relevant data for our readers. Conferences and sessions were held from time to time between the editorial board and the contributing authors to present the data in the most comprehensible form. The editorial team has worked tirelessly to provide valuable and valid information to help people across the globe.

Every chapter published in this book has been scrutinized by our experts. Their significance has been extensively debated. The topics covered herein carry significant findings which will fuel the growth of the discipline. They may even be implemented as practical applications or may be referred to as a beginning point for another development. Chapters in this book were first published by InTech; hereby published with permission under the Creative Commons Attribution License or equivalent.

The editorial board has been involved in producing this book since its inception. They have spent rigorous hours researching and exploring the diverse topics which have resulted in the successful publishing of this book. They have passed on their knowledge of decades through this book. To expedite this challenging task, the publisher supported the team at every step. A small team of assistant editors was also appointed to further simplify the editing procedure and attain best results for the readers.

Our editorial team has been hand-picked from every corner of the world. Their multi-ethnicity adds dynamic inputs to the discussions which result in innovative outcomes. These outcomes are then further discussed with the researchers and contributors who give their valuable feedback and opinion regarding the same. The feedback is then collaborated with the researches and they are edited in a comprehensive manner to aid the understanding of the subject.

Apart from the editorial board, the designing team has also invested a significant amount of their time in understanding the subject and creating the most relevant covers. They scrutinized every image to scout for the most suitable representation of the subject and create an appropriate cover for the book.

The publishing team has been involved in this book since its early stages. They were actively engaged in every process, be it collecting the data, connecting with the contributors or procuring relevant information. The team has been an ardent support to the editorial, designing and production team. Their endless efforts to recruit the best for this project, has resulted in the accomplishment of this book. They are a veteran in the field of academics and their pool of knowledge is as vast as their experience in printing. Their expertise and guidance has proved useful at every step. Their uncompromising quality standards have made this book an exceptional effort. Their encouragement from time to time has been an inspiration for everyone.

The publisher and the editorial board hope that this book will prove to be a valuable piece of knowledge for researchers, students, practitioners and scholars across the globe.

List of Contributors

Yiming (Kevin) Rong
Tsinghua University, China
Worcester Polytechnic Institute, USA

Hui Wang
Tsinghua University, China

Yi Zheng
Worcester Polytechnic Institute, USA

Maria Nienartowicz and Tomasz Strek
Poznan University of Technology, Institute of Applied Mechanics, Poland

Kumaran Kadirgama, Mustafizur Rahman and Rosli Abu Bakar
University Malaysia Pahang, Malaysia

Bashir Mohamad
University Tenaga Nasional, Malaysia

Farhad Nabhani, Temilade Ladokun and Vahid Askari
Teesside University, School of Science and Engineering, Middlesbrough, TS1 3BA, UK

Fuhong Dai and Hao Li
Center for Composite Materials and Structures, Harbin Institute of Technology, China

Xiaocong He
Kunming University of Science and Technology, PR China

Nicola Bianchi, Massimo Barcaro and Silverio Bolognani
Department of Electrical Engineering, University of Padova, Italy

M. J. Jackson, L. J. Hyde, G. M. Robinson and W. Ahmed
Center for Advanced Manufacturing, College of Technology, Purdue University, West Lafayette, IN, USA

Philip W. Loveday and Craig S. Long
CSIR Material Science and Manufacturing, South Africa

Paul D. Wilcox
University of Bristol, United Kingdom

Ying Xie
Harbin University of Science and Technology, China

Printed in the USA
CPSIA information can be obtained
at www.ICGtesting.com
JSHW011412221024
72173JS00003B/525